Why Space Will Freak You Out

The Scariest, Strangest Parts of the Universe

By Space Experts Dr. Kimberly K. Arcand and Megan Watzke

Illustrations by Robert Ball

"Nothing in life is to be feared. It is only to be understood."

—Marie Curie

The full color art was created using pencil, colored ink, and finished in Photoshop.

Published by Sourcebooks eXplore, an imprint of Sourcebooks Kids
1935 Brookdale RD, Naperville, IL 60563-2773
(630) 961-3900
sourcebooks.com

Cataloging-in-Publication Data is on file with the Library of Congress.

Source of Production: Wing King Tong Paper Products Co. Ltd.,
Shenzhen, Guangdong Province, China
Date of Production: August 2025
Run Number: 5050483

Printed and bound in China.
WKT 10 9 8 7 6 5 4 3 2 1

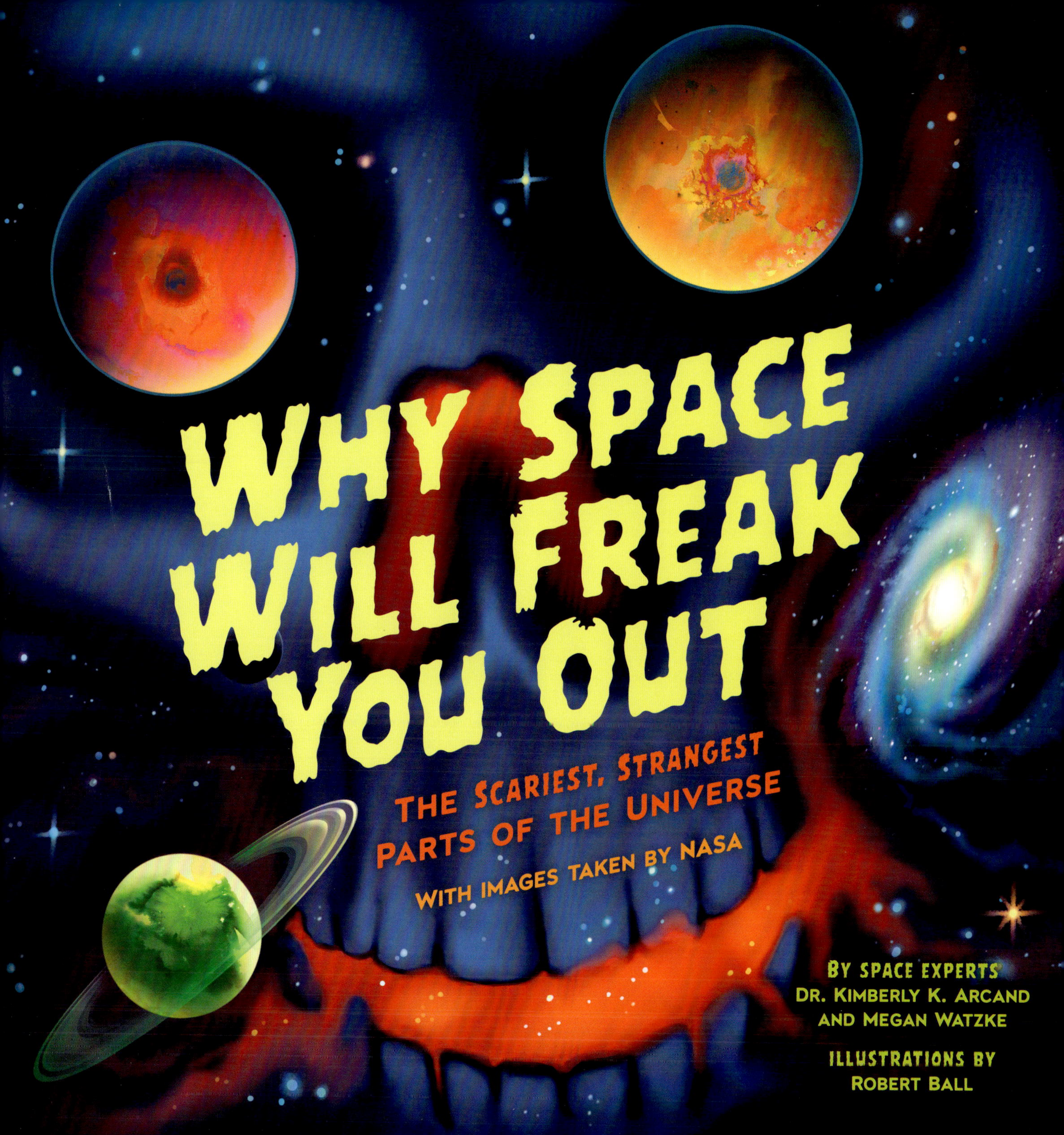

WHY SPACE WILL FREAK YOU OUT
THE SCARIEST, STRANGEST PARTS OF THE UNIVERSE
WITH IMAGES TAKEN BY NASA
BY SPACE EXPERTS
DR. KIMBERLY K. ARCAND
AND MEGAN WATZKE
ILLUSTRATIONS BY
ROBERT BALL

CONTENTS

Whenever you see a **bold word** in the book, flip to the glossary on page 108 to learn more!

The Most Dangerous Places in Our Galaxy

What Lies Beyond

WELCOME TO THE SOLAR SYSTEM—AND BEWARE!

WHAT DO YOU KNOW ABOUT OUR SOLAR SYSTEM? You probably know there are eight **planets** that orbit the Sun, along with an asteroid belt, comets, moons, and many smaller pieces of ice and rock farther out. Do you ever dream of hopping on a spaceship and venturing out into space to see some of these objects? Space travel is becoming more popular these days, but we're here to tell you, space is no joke!

Our solar system is full of freaky, scary, dangerous places. There are places nearby where there's not a drop—or even a molecule—of water to be found. Meanwhile there are worlds where liquids—including water!—flow, but are either deadly to reach or mixed with poisons.

Starting with our nearest star, the **Sun** itself, we should proceed with caution. What looks like a calm yellow ball of warmth from the Earth's surface is actually a boiling engine of energy and gas that sends dangerous blasts of radiation out into space. Thankfully, a cozy protective layer—our atmosphere—keeps us pretty safe here on Earth.

Apart from Earth, the planets in the Solar System are too exotic to live on for more than a few seconds without the safety of a spaceship or space suit. The two closest to the Sun, Mercury and Venus, host smorgasbords of dangerous outcomes, ranging from being baked to being crushed. Jupiter and Saturn are two gargantuan planets made up almost completely of gas, and beyond them lie the icy giants, including Neptune, where cold takes on a new meaning.

Some of the moons around Jupiter and Saturn may seem like they might be more hospitable at first glance. While a couple have conditions that mimic Earth, like oceans of water, mountains, and valleys, do not be fooled. Saturn's moon, Titan, has lakes and air that will suffocate a human instantly, while another of its moons, Enceladus, would trap you under miles of thick ice sheets in a salty tomb.

In short, there are places in our Solar System you don't want to go. Some are colder than our definition of *frozen*, and others will roast you at temperatures hotter than the Sun. You can be crushed to death or blown to bits in planet-sized storms. Ferocious volcanoes dominate worlds, and ice sheets plunge deeper than the deepest parts of our ocean.

Explore at your own risk!

MERCURY: THE TERRIFYING TWO-FACED PLANET

WELCOME TO THE FIRST PLANET that will freak you out: Mercury. Mercury is a planet of scary extremes. There's about a 1,100-degree difference between the freezing temperatures on one side of the planet and scorching temperatures on the other...and they're happening at the same time!

As the closest planet to our Sun, Mercury has an average distance of about thirty-six million miles away from the Sun, which is close! Couple Mercury's close distance to the Sun with its lack of an atmosphere, and you get a two-faced world of suffering.

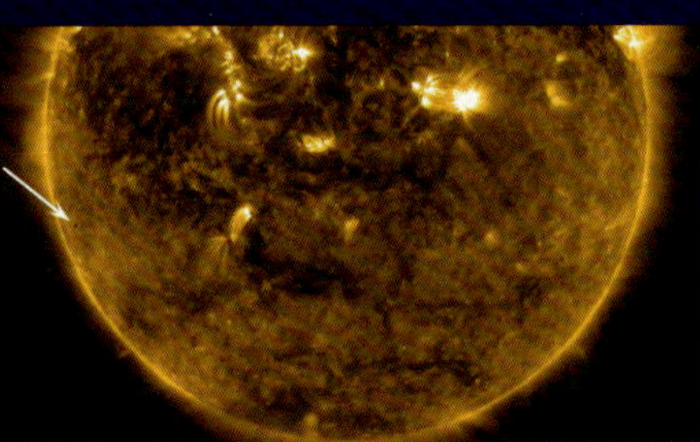

Mercury looks like the tiniest blip next to our Sun. It can be seen here as the black dot at the end of the arrow on the left side. This image captures Mercury just as it passes directly between the Sun and Earth, offering a convenient scale comparison for the smallest planet in our Solar System.

On the side of Mercury facing the Sun, the temperature on the surface reaches 800°F. That would be like sitting in your maxed-out oven, and then turning the heat up *another* three hundred degrees. (Note that 800°F is hot enough to melt lead.)

But what gives Mercury two faces? What about the freezing side of the planet? Mercury spins at a much different rate than the Earth. On Earth, we experience a full day and night when our planet completes a spin on its axis over twenty-four hours. Mercury, on the other hand, spins much more sloooowly. In fact, it takes about one hundred and seventy-six Earth days for Mercury to completely spin around once. This means that while one side of the planet is facing the Sun for a very long time, the other is shielded from sunlight for just as long.

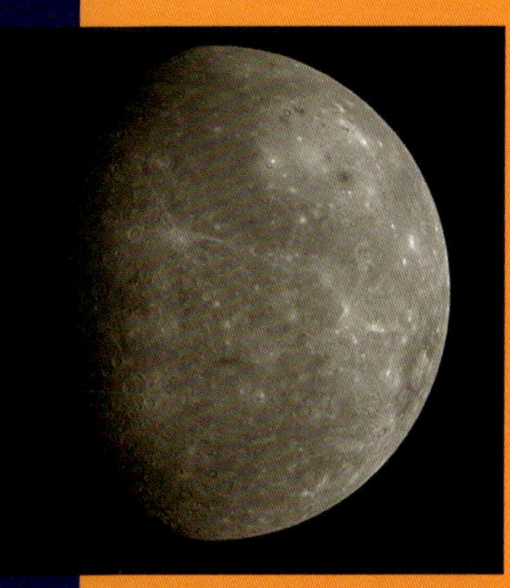

If you glance quickly at this image of Mercury, you might think it was our own Moon instead of the closest planet to the Sun. Mercury's surface has been ravaged by asteroids. Mercury has no atmosphere to speak of, a particular closeness with our Sun, and a very slow spin.

The side of Mercury facing away from the Sun—the shady side—is no comfortable place either. Because Mercury has virtually no atmosphere, the heat from the Sun almost entirely escapes into space. An atmosphere like we have on our planet, in addition to giving us air to breathe, acts like a blanket, trapping heat from the Sun underneath it. The thicker the "blanket," the warmer the surface of the planet can be.

With no way to keep the heat on the side facing away from the Sun, temperatures on Mercury dip as low as –290°F! What can survive at that temperature? Certainly no life like we know of, such as mammals, reptiles, or even microscopic life. So you'd be a frozen astropopsicle before you had the chance to say "Brrrr."

What does all of this mean? Well, it means Mercury has two very different real ways to kill you! Your best bet to not be instantly frozen or fried? Stay along the **terminator**, which is the line where day and night meet on Mercury. The region around there might have temperatures around –100°F. That's still colder than the coldest temperature ever recorded on Earth, however, so you can't stay there long unless you have a very protective space suit on.

If Mercury sounds like a terrible place, you're right. So far,

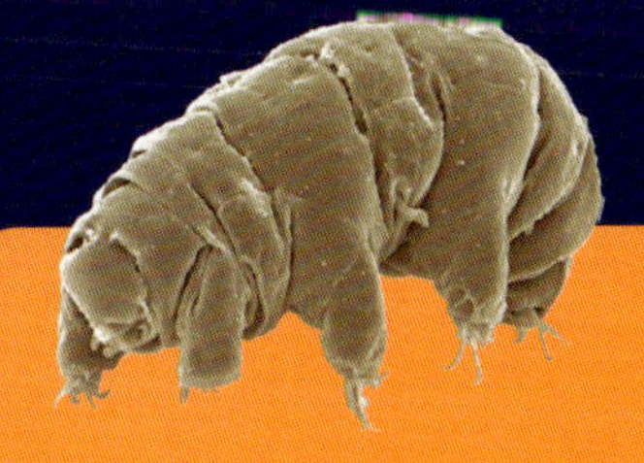

So far scientists have found the best survivalists that we know of are tardigrades, also called water bears, a tiny micro animal. But even tardigrades can't exist below –20°F, so they're not on Mercury. This is an image of a tardigrade under a microscope.

humans have yet to send anything to land on Mercury. However, they have sent spacecraft to go around the planet to learn more about it.

In 2011, NASA's MESSENGER (an acronym for *MErcury Surface, Space ENvironment, GEochemistry, Ranging*) spacecraft left Earth to become the first mission ever to orbit the planet. No humans can travel in such an orbiter, but such an uncrewed spacecraft is very important in giving us a close-up look.

With some tricks up its sleeve, like a titanium heat shield and other engineering marvels, MESSENGER survived this tortured environment in orbit for four years before crashing onto the planet's surface. Before its demise, MESSENGER sent invaluable information about the planet's interior and surface...and why none of us should try to book the same trip.

In a spacecraft, on the surface, or anywhere in between, the two-faced planet Mercury gives you a horrifying choice: bake to death on one side or freeze to death on the other. Our best advice is, steer clear!

Some of the pockmarked features on Mercury are striking, such as this collection of scars with deep shadows resembling Mickey Mouse, near the Magritte crater in the southern part of Mercury.

Since it doesn't have an atmosphere, Mercury has no weather to help erase the asteroid bombardment that happened during the earlier days of the Solar System. This has left Mercury scarred with tens of thousands of craters on its surface.

Mercury is only a little bit bigger than Earth's Moon, making it the smallest planet in our Solar System. If the Earth was a baseball, Mercury would be about the size of a golf ball.

At Mercury's distance, the Sun appears more than twice as big and ten times brighter in the sky than it does here on Earth.

Mercury is the fastest-moving planet in our Solar System, completing one "year," or trip around the Sun, in just eighty-eight Earth days. Because of this, people named Mercury after the Roman messenger god who was supposed to be pretty fast on his feet (and wings).

THE GOLDILOCKS ZONE

While it sounds like something of a fairy tale, the **Goldilocks zone** is an actual scientific term. It also goes by the name of the **habitable zone**. It refers to the area around a star where the temperature is "just right" to allow liquid water to exist on a planet's surface. In our Solar System, the only planet in the Goldilocks zone is Earth.

Closer-in planets like Mercury and Venus are too hot, and water (if it existed there) would immediately evaporate. And farther-out planets like Mars are too cold. Astronomers have found many planets in other star systems that are located in Goldilocks, or habitable, zones, although that does not necessarily mean liquid water is there. Since liquid water is so important to life here on Earth, finding it on other planets means life might exist there.

VENUS:
A CHOKING, CRUSHING, AND ROASTING PLANET

SCIENTISTS OFTEN CALL VENUS a sister planet to Earth. But that description works only if that sister is an evil sibling that chokes you to death. Yes, the two planets, Earth and Venus, are roughly the same size. Each also weighs about the same, and they seem to have a similar structure on their insides. And like Earth, Venus also has mountains, volcanoes, and flat plains. This is where the similarities end, however. There are many treacherous elements to Venus, including volcanoes run amok, but let's keep our focus on the primary difference for now: the atmosphere.

The fact that Venus has an atmosphere probably sounds like a good thing. After all, we have an atmosphere here on Earth, and it lets us breathe! But the atmosphere on Venus is a thick envelope of dense clouds and poisonous gas that provides a trifecta of ways to kill you: being choked, crushed, or roasted to death. Let's look at each of these suffer-fests separately.

Venus's atmosphere might look

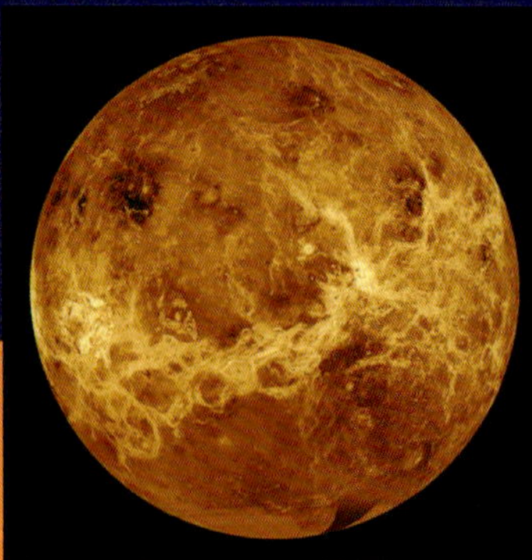

Left: Venus looks relatively peaceful with its fluffy burnt-marshmallow-like atmosphere. But it's actually a superhot planet with clouds of acid and a surface saturated with lava. Right: NASA's Magellan spacecraft captured this data of Venus, peering down through its cloudy atmosphere to the scarred, active surface of the planet.

Illustration of the three main layers making up Venus and Earth. If Venus and Earth were cakes and you could slice through them, they would look similar. Their layers would be like a rocky, crunchy crust, a solid-ish but still a bit squishy middle mantle, and a very hot mostly ooey-gooey center, or core.

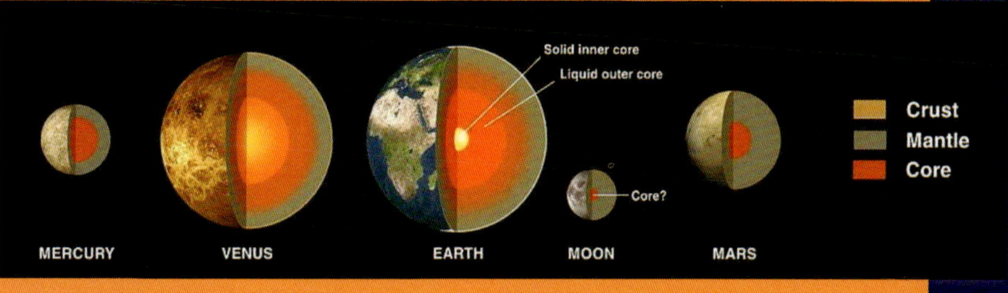

like a peaceful cloudy place from the outside, but it is made up mostly of carbon dioxide and sulfur acid. Humans can only exhale—not inhale—carbon dioxide, and sulfur acid will burn your lungs and stomach if swallowed or inhaled. Don't try either!

The atmosphere on Venus is also so thick that it will crush you, with a pressure on the surface about ninety times greater than on Earth. To reach that pressure on our planet, you would have to travel about three thousand feet down into the ocean. Or think about it this way: if you could stand on the surface of Venus, it would feel like having the weight of a small car on every inch of your body.

Finally, Venus's thick atmosphere makes it worse than being smothered under a hundred blankets on the hottest day of the year. The atmosphere traps so much of the heat from our Sun that the temperatures on the surface of Venus skyrocket to over 880°F. This makes Venus the winner for being the hottest place in our Solar System.

For the reasons above, to explore Venus, we have to send robotic missions if we want to get a closer look. For example, the European Space Agency launched Venus Express in 2005 to study the atmosphere from above. It managed to survive orbiting the planet for nine years, collecting data as it went.

Active volcanoes are shaping Venus's surface, as shown in this artist's illustration of a part of the planet's southern hemisphere, with the blotted Sun blazing in the distance.

NASA has had some of its spacecraft fly by Venus, but it is planning something major for the late 2020s. It's hoping that its DAVINCI (Deep Atmosphere Venus Investigation of Noble Gases, Chemistry, and Imaging) mission can drop a probe that makes it to the surface of Venus. At least for a little while. If all goes well, the probe will take pictures as it does its one-way descent, and we will all get a front-row look at this planet that none of us should ever go to ourselves!

So don't let Venus fool you. It may have lots of aspects reminiscent of Earth, but everything about this planet is dangerous if not lethal. Again, if Venus is our sister planet, it is an evil sibling, and we should consider staying far, far away from this branch of the family tree.

Venus isn't the closest planet to our Sun, at only the number-two spot, but it is the hottest—even hotter than Mercury—thanks to its thick atmosphere and volcanoes. This artist's illustration shows the eight planets of our Solar System along with some other bodies, like dwarf planet Pluto.

Venus spins in the opposite direction than every other planet in the Solar System. This could be a sign that Venus was hit by a rock or baby planet while it was forming about 4.5 billion years ago that sent it spinning in the opposite direction.

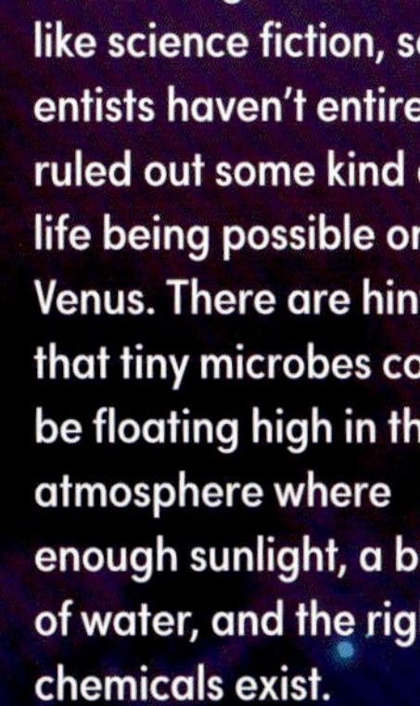

Even though it sounds like science fiction, scientists haven't entirely ruled out some kind of life being possible on Venus. There are hints that tiny microbes could be floating high in the atmosphere where enough sunlight, a bit of water, and the right chemicals exist.

Venus gets its name from the Roman goddess of love and beauty. In fact, most of the features on Venus have been named after women.

LEARN LIFT

THE RUNAWAY GREENHOUSE EFFECT

One of the biggest problems we face on our planet is climate change and the effects of what is commonly referred to as the **greenhouse effect**. As humans pump more gas like carbon dioxide and methane into the Earth's atmosphere, mainly through the burning of fossil fuels, we trap more and more of the heat from the Sun onto our surface and warm the planet.

The same effect has happened on Venus but to a much more advanced degree. Scientists think that Venus may have liquid water that could have been habitable by some sort of life we recognize about a billion years ago, but that is all long gone. The runaway greenhouse effect on Venus caused whatever liquid water that did exist to evaporate into the atmosphere. This added water vapor in the atmosphere trapped even more heat, and the cycle continued until the surface of Venus became a barren wasteland. Many scientists think this is an extreme version of what is happening to our planet now, and we can learn about Earth by studying Venus and its atmosphere.

JUPITER:
THE STORMIEST WORLD IN THE SOLAR SYSTEM

STORMS ON EARTH CAN BE PRETTY SCARY. But there is a storm on Jupiter that makes all the storms on our planet seem like a breezy day at the park.

The Great Red Spot is the mother of all storms. It's so enormous that the entire Earth could fit inside of it. From space, Jupiter's Great Red Spot looks like a piece of abstract art, filled with lines and swirls of color. But don't let that pretty face fool you. This is a vortex unlike anything you have ever seen!

The Great Red Spot is a much bigger and scarier cousin to hurricanes and typhoons on Earth. Hurricanes here will last a few days or maybe weeks. The Great Red Spot has been going on for *centuries*. This planet-sized storm has been raging for so long that we don't even know when it started. We can trace it back to at least five hundred years ago. But it could have been going on much longer, since we need telescopes to see it, and they weren't invented until the seventeenth century.

Not only has the storm lasted much longer, the

Jupiter is the fifth planet from our Sun and also the largest in the entire Solar System, more than twice as massive as the other planets combined. About a thousand Earths would fit inside Jupiter. In this image, can you see where one of Jupiter's moons, Europa, is casting the shadow on the planet? (See the dark spot on the lower left.)

A giant storm called the Great Red Spot blisters through Jupiter's atmosphere, creating these spiral-looking waves behind it.

experience inside the Great Red Spot would be much worse. The winds are at least twice as powerful as an Earth-based hurricane—howling around 250 miles per hour—and could rip apart the strongest buildings or structures here on Earth. Then there are the ferocious bursts of lightning and thunder. Finally, inside the Great Red Spot, only a fraction of sunlight cuts through, creating a tomb of darkness, wind, and electricity.

Scientists aren't quite sure what is making the Great Red Spot happen. One clue is that the Great Red Spot is an **anticyclone**, meaning that it rotates counterclockwise, which is abnormal. One theory is there are two gigantic streams of air flowing constantly feeding this beast. Think of giant conveyor belts moving in opposite directions but stacked on top of each other...and the size of a planet. Another theory is there is a gigantic planetary air vent, channeling warm air from deep down in Jupiter, that fuels the storm.

The truth is we don't know enough about what is inside Jupiter to understand this storm. NASA sent the Galileo probe into Jupiter in 1995, and it made it about 112 miles below the surface before it stopped sending signals, presumably crushed by the pressure of the planet's atmosphere.

That's another way that Jupiter would be the end of you: as you travel even just a tiny fraction of the distance toward its center, the pressure and temperatures go way up. Scientists think that if you could make it to Jupiter's core, the pressure would be about 650 million pounds per square inch of your body. Compare that to the very bottom of the Pacific Ocean, which has pressure of about sixteen thousand pounds per square inch. Humans cannot survive there without a very strong submarine!

Scientists have also learned that the pretty stripes and swirls that we see from the outside of Jupiter are lethal combinations of **ammonia** and clouds swimming in an atmosphere of hydrogen and helium, which would be dangerous to inhale.

In short, the Great Red Spot is one gigantic storm that doesn't want to go away. Even though they don't know exactly what is causing it, scientists agree that at least one of the reasons it's lasting so long is that Jupiter is mostly, if not entirely, made of gas. On Earth, our storms change and eventually fall apart when they hit land. Without any surface on Jupiter, there's no friction to slow or stop this cosmic hurricane of epic proportions.

It's not only the Great Red Spot that you'd have to be concerned about if you ever visit Jupiter. This image, using infrared light, shows stormy cyclone patterns across Jupiter's north pole. There is one main central cyclone storm surrounded by eight smaller cyclones with sizes up to 2,900 miles (4,600 kilometers) across.

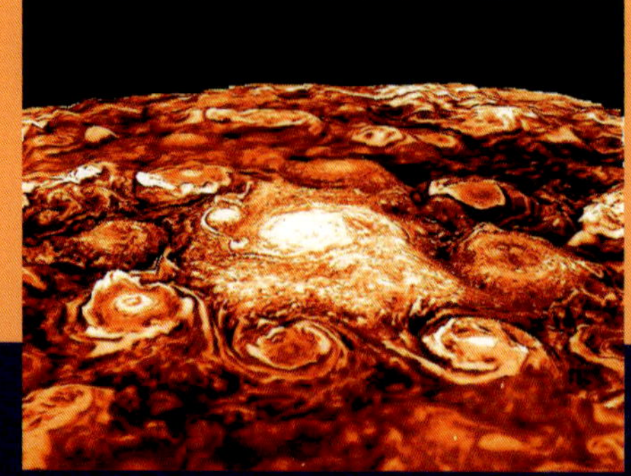

BRAIN BOOST

Jupiter is the fifth planet from the Sun in our Solar System and by far the biggest. About one thousand Earths could fit inside Jupiter. To put it another way, Jupiter is twice as big as all of the other planets in the Solar System combined.

NASA has sent several spacecraft over the past few decades to look at Jupiter. The Pioneer 10 and 11 spacecrafts and Voyager 1 and 2 missions, which all launched in the 1970s, flew past Jupiter. At the end of the twentieth century and the beginning of the twenty-first, the Galileo and Cassini-Huygens missions both orbited Jupiter for years before they stopped working!

Scientists still do not know what lies at the center of Jupiter. Some argue that the entire planet is made of gas, that is, hydrogen and helium. Others think that there may be a rocky core about the size of the Earth buried at its very center. If there is a surface down there, it would be over 40,000°F.

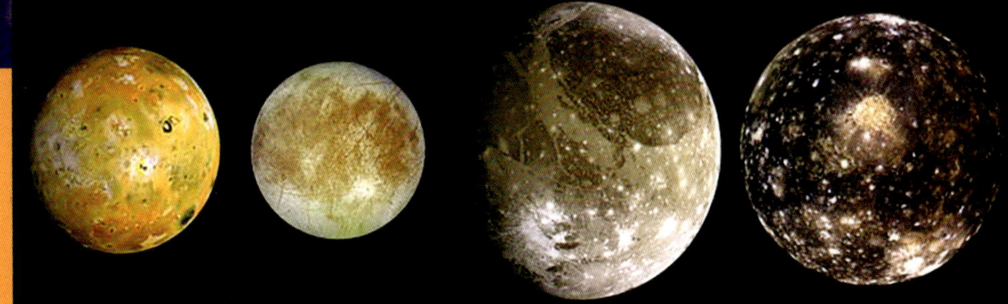

This collage of images shows Jupiter's four largest moons, which were first seen in 1610. From left to right, ordered by distance, are Io (closest to Jupiter), Europa, Ganymede, and Callisto (farthest from Jupiter). Io is the most volcanically active moon or planet in our Solar System. Europa has a solid rock/iron core with an ice layer at its surface. Ganymede is the most massive moon of Jupiter and in the entire Solar System. Callisto is very heavily cratered and pockmarked.

JUPITER'S MOONS

When the Italian astronomer Galileo Galilei first made his telescope and looked up at Jupiter, one of the first things he noticed were four dots of light next to it. These were Jupiter's four largest moons: Ganymede, Callisto, Io, and Europa. These names were chosen because they have some connection in mythology to Jupiter, the king.

Today, we know that Jupiter has many more moons than this quartet. At last count, Jupiter officially has ninety-five moons. Jupiter is sometimes called a **failed star**. Stars form when a really big cloud of gas and dust gets pulled together by gravity. As the cloud pulls in more debris, the center starts to get warmer. If this keeps happening for long enough, and the center of the cloud gets hot enough, then atoms and molecules begin to smash together. This is the power source for stars.

Jupiter is really big but not big enough to get this engine of a star going. It would need to be about seventy-five times bigger to light that atom-smashing process, or **nuclear fusion**, and become a second star in our Solar System. How weird would Jupiter have been had it been bigger! We would have two stars in our Solar System instead of only the one star—the Sun—we have.

IO:
THE KILLER ROTTEN-EGG MOON

HAVE YOU EVER GOTTEN A WHIFF of rotten eggs? Maybe you've been walking down a city street and had a nose full of the stench from a sewer. It's nasty! Now imagine an entire world that is enveloped in this putrid smell. This place exists! Not only is it foul-smelling, it can also kill you in a bunch of terrible ways.

Jupiter is our Solar System's biggest planet and has ninety-five moons. This nasty place is one of those moons, and its name is Io.

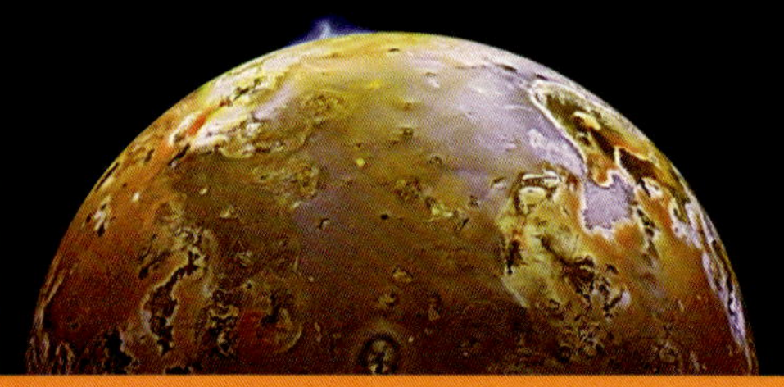

NASA's New Horizons mission zipped past Jupiter just in time to catch a volcanic eruption on Io, Jupiter's third-largest moon. This massive plume on Io is two hundred miles high, spewing out noxious chemicals in the moon's northern hemisphere.

One of the first discoveries Galileo Galilei made with his new telescope in 1610 were four dots next to the planet Jupiter. They were not stars, but moons. He called the brightest one Jupiter I. In the mid-1800s, other scientists renamed it Io after the Greek god Zeus's mistress, whom he later turned into a cow! This was long before scientists realized that Io smelled like a vat of rotten eggs.

Astronomers have sent spacecraft into the outer parts

of the Solar System for decades, hoping to find out what's happening in its darkest corners. So, in the 1970s, NASA sent its Pioneer and Voyager missions toward Jupiter and its moons, and the Galileo spacecraft passed Io multiple times around the turn of the twenty-first century. More recently, NASA's New Horizons and Juno missions have been in the vicinity of this smelly moon, sending back detailed photos of what is happening in this freaky place.

An artist combined NASA images of Jupiter and the four large moons discovered by Galileo: Callisto, Europa, Io, and Ganymede, along with a spacecraft into one illustration of the system. Jupiter is much larger than Earth; over one thousand Earths would fit in Jupiter. So Io, at about the same size as Earth's Moon, is much smaller in comparison to the planet it rotates around.

Jupiter has four pretty big moons and the rest are much smaller. About the size of Earth's Moon, Io sits in orbit around gigantic Jupiter and in between two of its big moons, Europa and Ganymede. These big planetary bodies have powerful gravity that pull and stretch Io from side to side. This back-and-forth generates extreme heat inside Io. In other words, Io is being squished and squeezed like a stress ball.

This movement generates heat, which needs somewhere to go. Cracks in Io's crust on the surface constantly stream rivers of lava to relieve this heat and pressure. Imagine walking along and pools of lava start bubbling along the ground. Think of the game *The Floor Is Lava* but in real life and across an entire world. And worse!

While those breaks and fractures in the surface release some of the heat right below Io's ground, they are not enough to do the job. This moon-in-a-gravitational-squeezer situation needs bigger release valves. To do that, Io has formed hundreds of volcanoes across the moon that erupt and spew lava. Io is the most volcanically active place in the Solar System.

Imagine yourself on the surface of Io. After navigating the danger of flooding lava, you'd need to constantly look for lava bombs from above as those volcanoes shoot lava hundreds of miles into the sky.

If that doesn't sound terrifying enough, let's return to the smell. Io's many volcanoes don't send out the typical dust and gas that Earth's volcanoes do. Instead, Io's volcanoes pump out toxic clouds of sulfur dioxide and other noxious chemicals. The air is completely poisonous and will burn your lungs from the inside out.

If you inhaled it, the sulfur dioxide would mix with the mucus membranes inside your lungs, turning it into sulfurous acid, which would suffocate you. The "good" news is that Io's atmosphere doesn't contain oxygen, so you'd probably lose consciousness within a matter of seconds before you feel the full effects of the sulfur dioxide. And you thought the rotten egg smell was bad.

As an artist has drawn here, Jupiter's moon Io is so volcanically active it might have a massive ocean of magma just under its entire surface, covering the whole moon! The volcanoes on Io are so busy that they frequently resurface the moon.

BRAIN BOOST

Io is the fifth moon in order away from Jupiter and the third largest moon.

Io is tidally locked, meaning the same side always faces toward the planet it orbits. This is true of Earth's Moon too.

Several NASA spacecraft have flown close to Io, including Voyager in the 1970s, Cassini in the 2010s, and the Juno mission recently swung by.

Io cuts across Jupiter's powerful magnetic fields and generates electricity while in orbit. Io generates as much wattage as one thousand nuclear power plants.

An artist has drawn an illustration of one exomoon candidate called Kepler-1625b-i, pictured here with its parent planet and star in the system behind it. If confirmed, Kepler-1625b-i would be the first moon to be found outside of our Solar System.

REACH FOR THE (EXO)MOON

As people have explored the Solar System, they have discovered that some of the moons around other planets, like Io around Jupiter, are quite fascinating. It's no surprise that as soon as they were able, scientists wanted to look for moons near planets and other stars as well.

Since the 1990s, astronomers have found planets around stars other than our Sun. These planets outside the Solar System are called **exoplanets**. It's very tricky to find exoplanets and it's even more challenging to find moons that orbit them. **Exomoons** are even fainter than the faint planets they orbit, but that hasn't stopped scientists from developing new instruments to give them a better look.

Scientists use the science of shadows to try to find exoplanets and exomoons, paying close attention to how light dips and changes as planets or their moons orbit around stars far away. So far, scientists have discovered a handful of exomoon candidates—mostly large Neptune-sized moons—and they hope that the next generation of telescopes will be even better at finding smaller moons and getting information about them. Who knows what they will find!

SATURN:
THE FALLING-INTO-OBLIVION PLANET

ASK A GROUP OF PEOPLE WHAT their favorite planet is besides Earth and chances are Saturn is near the top. After all, it looks so fantastic with its rings around the middle. If you ever have a chance to see Saturn through a telescope, it looks like the cutest cutout from a cereal box or cartoon. Do not let these impressions of friendliness fool you! Saturn is a deadly mosh pit that will, among other things, allow you to fall into oblivion, never to be seen again.

Let's begin with those famous rings, which might look solid but are not. They are made up of millions or even billions of pieces of ice and rock. Many of these pieces are composed of dust grains that are going at such high speeds, they can penetrate metal. Some of the rings contain chunks as big as houses, and a few are likely the size of mountains.

Saturn is a beautiful jewel of a planet in our Solar System with its dramatic rings as shown in this image from NASA's Cassini spacecraft. But don't be fooled by its beauty! It's also a death trap between the speeding particles in the rings and its pressurized atmosphere. Not to mention, lacking a surface!

To best avoid these deadly obstacles, you may think about approaching Saturn near the top or bottom. This is not a good idea either as Saturn has **auroras**. Earth's auroras are often referred to as the Northern Lights. On Saturn, where the magnetic field is much stronger than

Earth's, auroras are dangerous. They carry high-voltage electric storms that would fry any spacecraft's electrical and navigation systems.

On Saturn's south pole, an eerie glow of aurora is streaking out about six hundred miles from the tops of its clouds. Auroras are created by atomic collisions that create glowing gases.

Even if you could penetrate the surface of Saturn, the chances of landing anywhere are, well, zero. That's because Saturn is a giant ball of gas. (Scientists call Saturn and Jupiter **gas giants** for that very reason.) After the Sun formed about 4.5 billion years ago, the remaining dust and gas swirled around it and clumped together. Most of the lighter elements, like hydrogen and helium, got pushed outward, so Saturn and Jupiter are made mostly of these atoms.

The top of Saturn's atmosphere still has pools of hydrogen and helium, and the winds in the upper atmosphere are about four times stronger than Earth's most menacing hurricane. Past these, you'll find an icy world of ammonia crystals, which causes respiratory problems. This part of Saturn's atmosphere looks yellow because ammonia reflects sunlight and then brown due to other chemicals in the crystals. I don't know about you, but I wouldn't want to see only yellow and brown!

Closer in, things get worse. As you descend, the temperature and pressure keep going up, up, up. When you arrive at the core of Saturn, the temperatures skyrocket to about 15,000°F, which is hotter than the surface of the Sun. The pressure—or squeeze factor—at Saturn's core is so great that hydrogen, which is usually a gas, turns to liquid and starts acting like a metal, conducting electricity.

Again, don't be deceived by Saturn's beauty. Any spacecraft would be crushed, melted, and vaporized by the extreme pressures and temperatures almost as soon as it enters!

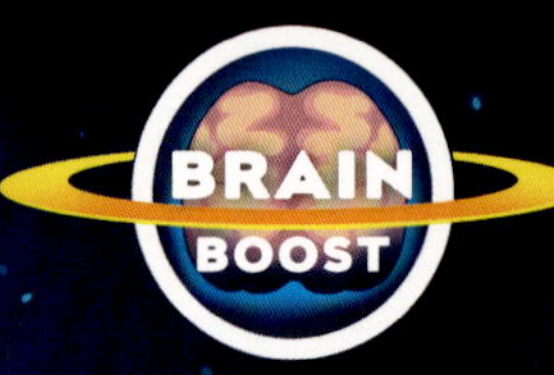

Saturn spins quickly and has a day of only 10.7 hours. This is the shortest day in the Solar System. Its year, however, is long because it is so far away from our Sun. It takes 29.4 Earth years for Saturn to complete one lap around the Sun.

Saturn is the largest planet in the Solar System except for Jupiter. With a diameter of about 72,400 miles, Saturn is about nine times wider than Earth. About 760 Earths could fit inside Saturn. To put it another way, if Earth were a nickel, Saturn would be the size of a volleyball.

Saturn is the most distant planet humans can see without a telescope, and it has been watched for millennia. When Galileo Galilei saw the rings of Saturn through his telescope in the seventeenth century, he called them handles, thinking they resembled what you grab to pick up a pot.

NASA's Cassini spacecraft plunged into the atmosphere of Saturn to end its mission after an incredible thirteen-year run around the ringed planet. In this image, one of Saturn's moons, Enceladus, is shown peeking through Saturn's upper atmosphere as Cassini begins its dive into the gas giant before being crushed and melted.

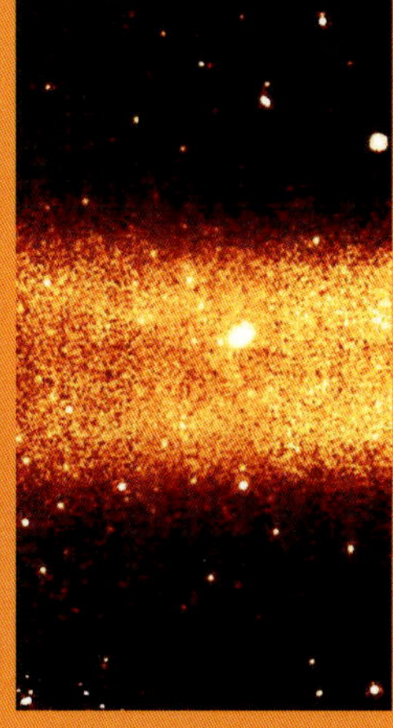

Saturn's rings are very thin overall and look razor sharp (left). Made up of ice, dust, and chunks of rock, a slice of Saturn's largest ring is highlighted in this image (right) as seen in infrared light by NASA's Spitzer Space Telescope. Spitzer was able to capture infrared light, or heat, from the dusty ring material swirling around the planet.

PUT A RING ON IT

Saturn's rings are definitely its most famous feature, but often misunderstood. It looks like the rings are solid, but they are made up of countless particles of different sizes. The rings are even more complicated than that. There are seven main rings, each orbiting at different speeds around the planet with space in between. The rings of Saturn are also extremely flat and thin. They stretch out about 175,000 miles away from Saturn but are only about thirty feet tall in most places.

Where did Saturn's rings come from? The jury is still out, but some scientists think the rings are the result of two of Saturn's moons colliding during the early days of the Solar System, creating the icy debris fields we see today.

Speaking of those debris fields, tiny particles of ice are currently streaming away from the rings toward the surface as the planet's magnetic fields slowly drain the rings. Scientists call this *ring rain.* The magnetic fields are pulling a couple of Olympic-sized swimming pools worth of ice from the rings every hour! At this rate, the rings will disappear in about three hundred million years.

TITAN: THE ANTIFREEZE AND FERTILIZER MOON

IF THERE IS ANOTHER PLACE IN THE SOLAR SYSTEM that looks like it would be a welcoming place to visit, Saturn's largest moon, Titan, might be it. Titan has clouds and lakes, rivers, and oceans. It even has a weather cycle in which precipitation falls onto the surface, pools into lakes and oceans, runs in rivers, and then evaporates into the atmosphere to start the process over again. This is very much how our weather system works here on Earth. Sounds like a great spot!

There's a catch. The earthlike features on Titan aren't filled with clean water and fresh air. Instead, the planet's air, land, and water are made up of **hydrocarbons**, a family of molecules that include crude oil, natural gas, antifreeze, and fertilizer.

Titan's atmosphere is a thick, hazy golden smog, which makes it hard to see through. Most of the atmosphere is made up of methane and ethane. Both of these molecules contribute to climate change here on Earth, but we have a lot more methane due to human production and cows that release it when they burp and fart (fact)!

These images of Titan were taken (from left to right) in October 2005, December 2005, and January 2006. Across those few months, you can see dramatic changes in its clouds: a brightening north and south pole in December, and in October and January, a bright region slightly below the middle nicknamed the "chevron." This may be a huge eruptive spot of carbon dioxide.

On Titan, the methane in its atmosphere is broken apart into hydrogen and carbon atoms when ultraviolet light from the Sun hits it. These atoms then join up with other atoms floating around to make other hydrocarbons that become part of the weather on Titan. Some scientists call Titan a chemical factory because it produces so many hydrocarbons using the methane in its atmosphere.

What might it be like to stand on Titan (with an oxygen pack on, of course)? This artist's illustration helps us imagine it! Think of dim daytime skies, jagged, icy rock formations, dark coffee-ground-looking dunes, and rivers and lakes flowing brown. You also might have a front-row seat to view Saturn and its gorgeous rings.

So if you wanted to find a planet with fossil fuels, Titan is the winner. Titan has hundreds of times more liquid hydrocarbons than all the reserves we know about on Earth combined. Even its sand dunes on the surface, which resemble coffee grounds, are made up of hydrocarbons. Just one lake on Titan would provide hydrocarbons for the entire United States' energy consumption for three hundred years, and scientists have found dozens of these lakes!

You may be wondering, why doesn't Titan blow up like a giant firecracker with all these fuels both above and below? The answer is that there is no oxygen on Titan, which you need for fire. Think about putting out a fire by covering it—you are essentially taking away the oxygen for the fire to burn. As long as you don't provide any oxygen, no giant explosion!

Of course, there are other issues with Titan. One major problem is the temperature. Scientists estimate that temperatures on the surface are around −290°F. This level of cold would freeze a human in about one minute. Then there are Titan's oceans. Many scientists are fascinated with Titan because there is evidence that it has oceans of water, which is fantastic for people looking for life outside our planet. That's the good news.

The bad news is that there is also probably ammonia in the oceans, which acts as an antifreeze and keeps the water from completely freezing over but makes the water hazardous to human health. Methane is also in Titan's seas locked inside of molecular cages called lattices. On Earth, we use liquid methane for rocket fuel, so imagine how dangerous that is!

Some scientists think **cryovolcanoes**—volcanoes that release chilled water instead of molten lava—at the bottom of Titan's oceans are continuously releasing methane, which is freed from the lattices once it reaches the atmosphere. This provides a fresh supply of methane that is constantly being split apart by light from our Sun.

Titan may seem to be our best bet if humans ever want to live somewhere else in the Solar System. But it may also be a place that blows us up, freezes us to death, or poisons everyone!

Titan, Saturn's largest moon, passes in front of Saturn and its dramatic rings. This image was taken by NASA's Cassini mission when it was about 1.4 million miles (2.3 million kilometers) away from Titan.

BRAIN BOOST

It takes sunlight about eighty minutes to reach Titan, and the sunlight that arrives there is about one hundred times fainter on Titan than here on Earth. This means it would never get brighter on Titan than twilight or dusk on our planet.

The atmosphere on Titan's surface is about 60 percent thicker than on Earth, which would feel similar to the bottom of a swimming pool. This means humans could walk around without a pressurized space suit, assuming they had oxygen and other life support.

Like the Earth's Moon, Titan is tidally locked, so the same face is always pointing toward Saturn. Titan is tilted like Earth, so it also has seasons. However, because it takes Titan and its parent planet Saturn twenty-nine years to complete one orbit around the Sun, its seasons are much longer, at about seven years per season.

The Cassini-Huygens probe spent a harrowing two and a half hours descending to the moon Titan before it landed safely on January 14, 2005, as illustrated in this artist's concept (left). It then spent over an hour sending data and images (right) of the murky rocky moon back to Earth before shutting down forever.

THE HUYGENS PROBE

The Cassini-Huygens spacecraft was a robotic spacecraft that NASA and the European Space Agency sent to study Saturn and Titan. After traveling for seven years within the Cassini orbiter to the Saturn system, the Huygens probe, which was nestled inside, went to Titan.

The Huygens probe was released from Cassini on December 24, 2004. It took three weeks before the probe entered the moon's atmosphere. After it blew off its heat shield, Huygens started snapping pictures. It sampled the atmosphere and measured electrical charges. It fell through clouds and haze, encountering turbulence along the way. It landed on something solid yet soft, similar to damp sand or wet snow. For seventy-two minutes, Huygens sat on the surface of Titan taking photos of its landing spot, which was littered with rocks and pebbles, and beaming them back to Earth, before its batteries ran out.

ENCELADUS:
THE SALTWATER TOMB

WHILE TITAN IS THE BIGGEST of Saturn's moons, Enceladus is one of its most intriguing...and dangerous. Enceladus does not look like anything else in our Solar System. It is bright white and covered in a thick layer of ice. In certain parts of the moon, there are these gigantic slashes, or tiger stripes as scientists call them, which may hold the key to a completely different world beneath the ice.

As you might expect for an entire planet entombed in ice, it's really, really cold on Enceladus. Because the surface is bright white, it reflects almost all the light from the Sun back into space, never warming the planet. On the surface, the temperature hovers around –330°F.

The icy, cratered, and fractured world of Saturn's moon Enceladus was captured during a flyby of NASA's Cassini mission. A famous set of tiger stripes, four large eighty-five-mile-long (135 km) fractures, are shown crossing the moon's south polar terrain. Scientists think that the blue-colored areas are where walls of the fractures in Enceladus's surface expose large pockets of coarse ice.

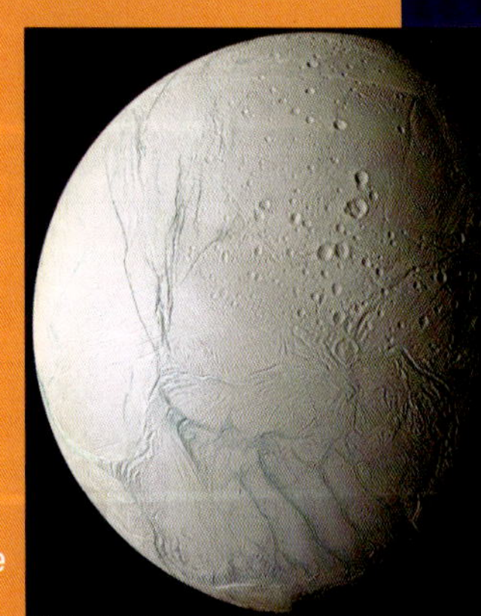

Also, the ice is not like any ice sheet we know about on Earth, unless you happened to be alive during the Ice Age a hundred thousand years ago. Even then, Enceladus is next level. At the thinnest part near the poles, the ice sheet is about three miles thick. Scientists think that everywhere else, the ice sheet could descend

Large, violent plumes of water, ice, and vapor are released from geysers along the tiger stripes of the southern part of Enceladus in this image from NASA's Cassini mission .

between twelve and sixteen miles down. In Antarctica, the deepest ice on our planet doesn't extend past about 1.5 miles.

You might think that Enceladus sounds like a frozen ball, so why bother with it? People are drawn to this distant moon not because of the cold and ice, but for what may be underneath it. Those tiger stripes are a clue. They show that the geology of Enceladus—that is how it's built—is changing.

Scientists believe that Enceladus has **tectonic plates**, related to what we have on Earth. These are large pieces of crust that can move and rub up against each other. When tectonic plates move on Earth, we get earthquakes. On Enceladus, cracks form in the ice, like when an asphalt road breaks up over time with extreme heat and cold.

The cracks on Enceladus have allowed about one hundred **geysers**, or erupting hot springs, to form on the surface of this moon. NASA's Cassini mission captured amazing images of these icy geysers that spray water out into space.

The instruments on these spacecraft can also identify chemicals in the liquid that Enceladus's geysers spew out. The water coming from within Enceladus has carbon, hydrogen, nitrogen, and oxygen—the molecules necessary to make life here on Earth. Any life that might exist on Enceladus, however, would be buried underneath those miles of ice, in what scientists think is an ocean of water that covers the entire moon.

You might wonder if the surface is beyond freezing cold, how is there possibly liquid water anywhere on the planet? There is a source of heat *inside* the moon. Enceladus orbits Saturn in between two other moons, Mimas and Tethys. Because of the push and pull from these other moons, and the gravitational force from its parent planet, Saturn, scientists have concluded the inside of Enceladus must be churning and generating heat.

The theory is this: There are vents in the bottom of Enceladus's planet-spanning ocean that pump heat into the water, keeping it liquid. These vents may also be releasing important minerals into the water that could make it a good environment for tiny microbes or other forms of life!

While humans haven't sent a probe to Enceladus before, scientists are working on it. The current idea from NASA is a mission that would launch in the late 2030s and arrive on Enceladus by the middle of the century.

Enceladus looks to be balancing dramatically on one of Saturn's rings in this image captured by NASA's Cassini mission. The scene looks eerie as it is backlit by our Sun. This lighting lets you see the icy particles coming from the geysers of Enceladus's south pole. Pandora, Saturn's tiny moon, which is only about fifty-two miles (eighty-four kilometers) wide, was opposite the rings when the image was taken (lower right).

Enceladus is a really small moon—only about one-seventh the diameter of Earth's Moon. At about 314 miles across, you travel more miles if you go from Los Angeles to San Francisco.

Tiger stripes are not everywhere on Enceladus. Some regions are nearly completely smooth, and others have craters that span up to twenty-two miles across. This diversity across the moon's surface leads to more questions about what's happening under the ice.

The geysers on Enceladus shoot their icy water out into space at about eight hundred miles per hour. Some of it ends up in Saturn's rings. The rest falls back to Enceladus in the form of snow, which is the only place other than Earth where that happens.

LOOKING FOR LIFE IN EXTREME PLACES

To learn more about the possibilities of life elsewhere, scientists have traveled to some of the most remote places on Earth. They drill giant holes to reach lakes under Antarctic glaciers to see what kinds of life exist there. They use special instruments and send robotic vehicles to the bottom of the ocean floor to learn more about **hydrothermal vents**, the same type of vents that could be keeping some of Saturn's moons warm enough to have liquid water. And they have looked at microbes, mainly bacteria and fungi, that survive in Earth's atmosphere despite extreme cold, very little or no water, and exposure to high levels of radiation.

Imagine finding extreme forms of life on Earth that can not only survive, but actually thrive, in some of the worst conditions on our own planet! Whether it's in the driest desert, in the deepest parts of the ocean, or under the thickest shelves of ice, scientists keep finding life. For example, there are alkaline lakes, which are called **soda lakes** because their **pH levels** are so high, that are filled with salts and ammonia. They have thriving ecosystems! From algae blooms to shrimp, and from specially adapting fish to various kinds of birds, soda lakes can support diverse kinds of life that adjust over long periods of time.

This type of Earth-bound research helps us better understand the possibilities of where life, in whatever form, might exist elsewhere in our Solar System. This is a whole field of science, called **astrobiology,** that combines what we know about Earth with what we are learning about in space.

Salt lakes on Earth offer places to study life in extreme environments. This image shows the Great Salt Lake in Utah, the largest saltwater lake in the western hemisphere. The water is saltier than seawater but still hosts many kinds of birds, shrimp, and other kinds of life, like phytoplankton and bacteria.

NEPTUNE:
A WINDCHILL FOR THE AGES

IF YOU LIVE OR HAVE VISITED some place where it gets below 50°F, you have probably heard of the term **windchill**. This is something meteorologists say when they are trying to explain how cold it feels when you combine a chilly temperature and gusts of wind.

Because the wind carries heat away from the human body, windchill is always lower than the temperature alone. Two Antarctic explorers, Paul Siple and Charles Passel, developed the first formula for windchill and coined the term in the midtwentieth century.

The record for the coldest windchill in the United States was logged on the top of Mt. Washington in New Hampshire at a frigid –108°F. On that day, the air temperature was –46°F, and there were wind gusts of 127 miles per hour. In parts of Antarctica, the air has been as cold as nearly –100°F, and with the winds there, the windchill drops to nearly –150°F. Brrr!

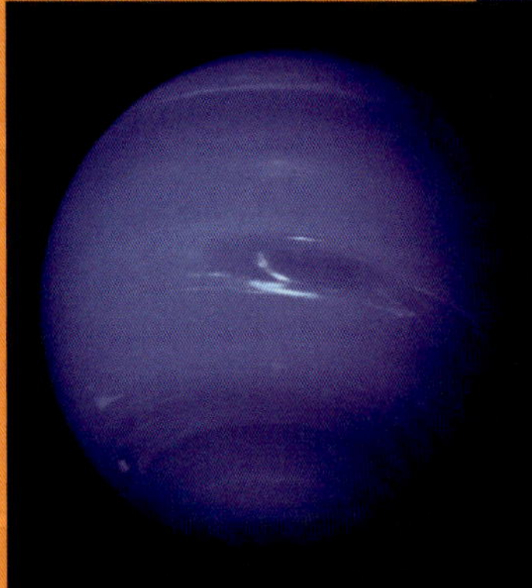

Neptune is a cold ice giant of a planet. This picture of Neptune was captured by the Voyager 2 mission when it was about 4.5 million miles away. Voyager also spied Neptune's Great Dark Spot (center) and a fast-moving bright feature called Scooter (lower left side).

You might think all you need are several coats to keep you warm, but the main problem with windchill is **frostbite**! Frostbite—where the tissue beneath the outer layer of skin freezes—can start to happen after a few minutes of skin exposure in –20 degree windchill. Frostbite can lead to fingers and toes turning black and even falling off or being amputated. Hypothermia is even more dangerous and happens when a body's temperature drops too low. Extreme cold can kill you!

Seven of Neptune's fourteen known moons—Galatea, Naiad, Thalassa, Despina, Proteus, Larissa, and Triton—appear in this infrared view from the Webb telescope. Triton dominates the image as a very bright point of light in the upper half. The green spikes around Triton are an artifact of the Webb camera.

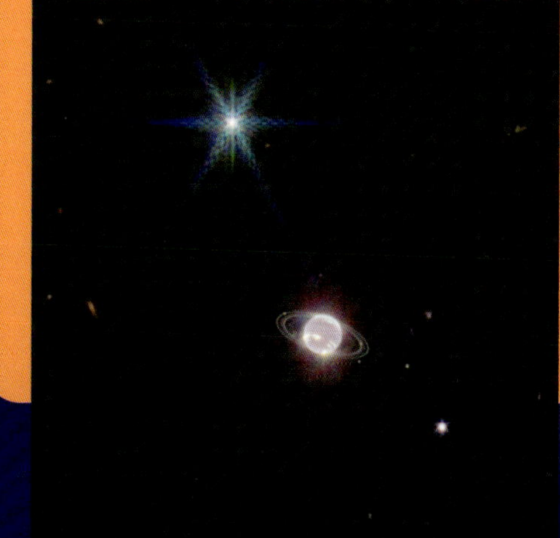

While windchills can be really scary on Earth, it's much worse on Neptune. Neptune is the eighth and officially most distant planet in our Solar System. It's very similar to Uranus, and scientists refer to both of them as **ice giants**.

With an average distance of nearly three billion miles away from the Sun, it takes sunlight four hours to reach Neptune, and it doesn't deliver much heat once it does. It's also the windiest place we know about in the Solar System. When Voyager 2 passed by in 1989, it clocked Neptune's winds at over one thousand miles per hour, and more recent estimates have the wind speeds howling at nearly 1,500 miles per hour!

And on this dark planet so far from the Sun, scientists estimate that the average temperature is around –350°F due to the gusts of wind. At the speeds Voyager and other telescopes have clocked, Neptune's winds are traveling about three times faster than commercial airliners and about twice as fast as the speed of sound!

One mystery surrounding this whipped and frozen planet is, where does this weird weather come from? On Earth, the weather is driven by heat delivered by sunlight, with more landing on the equator and less at the poles, for example. The differences in temperature over land and water cause the air to move and weather to form.

Neptune is too far away from the Sun and gets too little heat to be a source of its extreme weather. In order to generate the extreme blustery conditions on Neptune, scientists think there must be some sort of internal heat source from the planet's core.

Whatever lies beneath Neptune's pretty blue exterior remains mysterious, but you know anything down there is going to be colder than any cold we can ever experience here on Earth...and far more deadly!

Voyager 2 captured an up close view of Neptune's large moon Triton, the seventh-largest moon in the Solar System. This moon was likely a dwarf planet before being captured by Neptune's gravitational pull. Triton is also geologically active, like Earth, meaning the moon's surface and interior move and change over time. Canyons and mountains can form, volcanoes can erupt, and the surface evolves. In this image, you can see dark streaks produced by geysers on its icy surface.

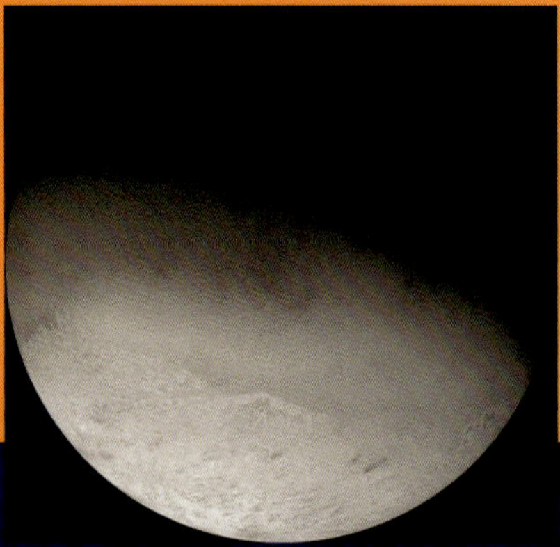

BRAIN BOOST

Neptune was the first planet discovered using math. In the nineteenth century, French astronomer Urbain Le Verrier predicted that Neptune should exist based on fluctuations he saw in Uranus's orbit. In 1846, German astronomer Johann Galle spotted it.

Neptune is named for the Roman god of the sea; it has six rings and at least thirteen moons. One of the moons, Triton, spouts geysers of icy plumes several miles above its surface.

Neptune consists of a fog of water, ammonia, and methane wrapped around a solid core about the size of the Earth. Above that is an atmosphere of hydrogen, helium, and more methane. It's the methane that gives Neptune its bright blue color because it absorbs the red part of sunlight and reflects the blue wavelengths back into space.

The closest humans have ever gotten to Neptune was the Voyager 2 mission that came within about three thousand miles of its north pole in 1989. We've never sent a spacecraft to orbit it or a probe to descend inside.

NASA's New Horizons spacecraft captured this spectacular view of Pluto, looking like it has a heart on its right when it was about 280,000 miles (450,000 kilometers) away. Pluto is an icy world full of exotic landscapes with towering mountains, huge ice sheets, pits, and valleys like nowhere else in the Solar System. Pluto lies in the Kuiper Belt, a section of icy bodies and other dwarf planets at the edge of the Solar System.

WHAT HAPPENED TO PLUTO?

In 2006, the global organization officially responsible for naming and classifying objects in space, known as the International Astronomical Union (IAU), decided there were certain characteristics that an object needed to be called a planet. By this new definition, Pluto no longer fit in with the other eight planets in the Solar System. Instead, the IAU decided that Pluto has more in common with thousands of small rocky bodies that orbit farther out from the Sun, known as the Kuiper Belt. That same year, in 2006, Pluto was officially recategorized as a **dwarf planet**, which means it's smaller than the other planets.

No matter what you call Pluto, however, it is still a fascinating world with mountains, craters, glaciers, and valleys. It also gives us a glimpse into the earliest days of the Solar System because scientists think it hasn't changed much over the last four billion years. We have already sent one spacecraft, called New Horizons, to explore it, and hopefully there will be more in the future.

THE MOST DANGEROUS PLACES IN OUR GALAXY

If you were hoping that space would be safe and relatively calm once we get out of our own Solar System, you may be disappointed. The Solar System is just one tiny piece of a much larger structure called the Milky Way **galaxy**. You can think of our Solar System like a neighborhood, and the galaxy is the whole city. This galactic city, however, isn't like any you've ever seen before.

Our Earth orbits the Sun, which is one star out of the three hundred billion or so stars in the Milky Way galaxy! The Milky Way is built like a giant pinwheel in space, with huge arms stretching and curving out from its center. Earth sits about two-thirds of the way toward the edge of one of the spiral arms.

We should be glad we are not near the energetic heart of the Milky Way. As it turns out, there are spine-tingling dangers lurking in practically every corner of our galaxy. In recent years,

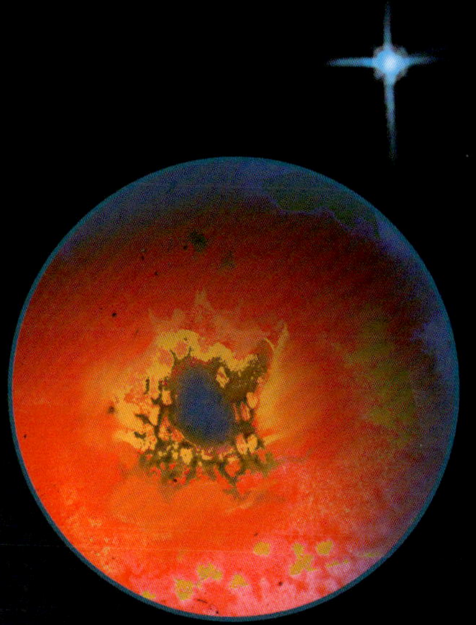

scientists have discovered thousands of planets outside of the Solar System. Astronomers call these exoplanets. If you thought Titan and Enceladus were wild, wait until you see what the rest of our galaxy has in store.

Astronomers have found exoplanets that are so creepy, they seem like they belong in a sci-fi show! There is an exoplanet that is completely cloaked in darkness, another where glass is raining sideways, and one where the entire surface is covered in lava. Our galaxy has planets that have come back from the dead, and another that's being devoured by the star it orbits.

It's not just exoplanets where our galaxy shows its bizarre and even ghastly side. The Milky Way has stars that explode and dead stars that send devastating blasts of energy out into space. And in the center of the Milky Way, a giant **black hole** that is millions of times the mass of the Sun lurks, waiting patiently for its next meal.

So our galaxy is no safe haven for the freakish things that exist in space. If you choose to leave our Solar System and keep exploring, you will find even more ways that space can send shivers down your spine!

POLTERGEIST:
A ZOMBIE PLANET AROUND AN UNDEAD STAR

THERE ARE FEW THINGS CREEPIER than zombies. After all, something that comes back from the dead is bloodcurdling. It should be alarming, then, that scientists have dubbed real-life planets as **zombie planets**, right? While these worlds are not covered with flesh-eating, virus-carrying monsters, they have their own horror stories to share.

Zombie planets are turned into these undead worlds not by their own doing, but because of the stars they orbit around. Stars have a finite amount of fuel to burn. This fuel is in the form of atoms that smash together within the inferno of the star's core. Everything is all well and good until, one day, the star runs out of those atoms to clobber together.

This artist's illustration shows what a creepy world like PSR B1257+12 b might look like. Also known as the Poltergeist planet, PSR B1257+12 b was one of the first and weirdest planets to be discovered beyond our Solar System. PSR B1257+12 b is a small rocky world that travels about twenty-five days to orbit its parent star.

There are several frightening paths that stars take at the end of their lives. If the star is large enough, it comes crashing down onto itself and is blown into smithereens in an event astronomers call a **supernova**. Such a mega explosion is disastrous not only for the star, but for any poor planets that happen to be circling around it. If our Sun were to explode as a supernova, the Earth and most of the other planets in the Solar System could be vaporized. Thankfully, our Sun won't explode because it's not big enough.

If a planet does manage to survive the onslaught from the supernova explosion, it may have to endure another wave of horrible things. Because the original star may not be done yet. Sometimes, the core of the collapsed star remains. However, the remaining star is no longer like it was in its previous life. Instead, it is a smaller, angrier, more powerful version of its former self known as a **neutron star**.

Pulsars and neutron stars are strange and extreme objects. Any spacecraft flying near one, like this pulsar illustrated by an artist, would be torn to pieces well before it reached the undead star. The magnetic fields around them are so strong, the force can pull spherical atoms into the shape a hot dog! The rapid rotation of a pulsar creates an incredible electricity generator that can make quadrillions of volts, which is thirty million times greater than those of lightning bolts!

If you are unlucky enough as a planet to a) survive a supernova explosion and then b) be circling around a neutron star, get ready. Neutron stars are some of the densest objects anywhere in the Universe. The violence of the collapsing star crushes every proton and electron into neutrons. Neutron stars are made up of mostly neutrons that can pack together tighter than anything else in the known Universe.

Neutron stars give off energy because they are born spinning incredibly fast. They also have superpowerful magnetic fields, also left over from the crash of the star. Combine these things, and you have a city-sized object packed with the entire mass of our Sun. This gives neutron stars an intense level of density with one teaspoon of its matter weighing about four billion tons, or about 8,000,000,000,000,000 (eight quadrillion) pounds.

Since density and mass are tied to gravity, neutron stars are the bully on the block as far as their gravitational power goes. Their gravity is about two billion times stronger than on Earth. This kind of gravity could tear a planet to pieces just through that force alone. More than that, neutron stars can blast intense waves of energy, zapping anything in their path. (Imagine a killer lighthouse!) Some neutron stars can spin hundreds of times a second, sending blistering bursts of radiation faster than you can blink your eye.

Despite all of this misery around these neutron stars, somehow, some way, there are planets that survive in their midst. These planets have lived through the apocalypse of a supernova and lasted through being neighbors with an aggressive neutron star. Scientists refer to these as zombie planets because it's like they have come back from the cosmic dead.

What would it be like to be on one of these freaky zombie planets? Not pleasant. Take the zombie planet called Poltergeist. Astronomers also call it by a much less exciting name: PSR B1257+12 c. Poltergeist was one of the first planets astronomers discovered outside of our own Solar System. About 2,300 **light-years** away, it orbits a spinning neutron star named Lich. The name Lich comes, appropriately, from a fictional undead creature.

While Lich is superhot, Poltergeist has no atmosphere because it's almost constantly getting blasted with radiation. Poltergeist also has no protection from this steady stream of radiation from Lich, which hits the surface and then bounces back into space. This means the surface of Poltergeist is incredibly cold. Of course, no atmosphere also means no air to breathe. Astronomers think that Poltergeist has two neighboring planets that are also in this same sad state. They've named them Phobetor and Draugr after the Greek god of nightmares and an old Scandinavian name for *ghost* or *undead*.

So with deadly radiation, guaranteed suffocation, and cosmically freezing temperatures, let's pencil Poltergeist in as one of the scariest places in space to visit—ever.

A massive star exploded and left behind this strange and beautiful supernova remnant called Puppis A. Shown in infrared light, it showcases data from NASA's Wide-field Infrared Survey Explorer.

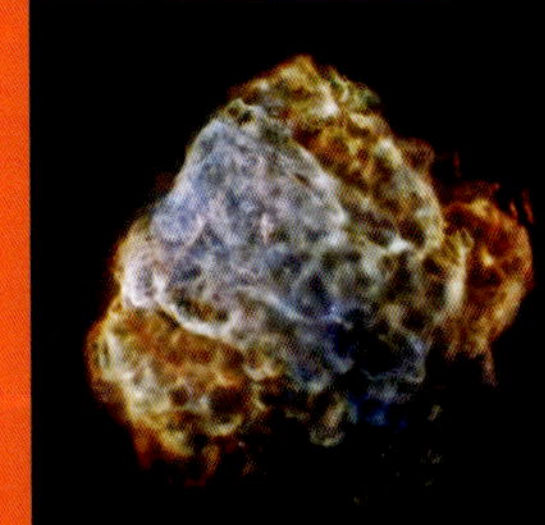

Because neutron stars spin consistently, some scientists have proposed using them for an **interstellar** GPS to help astronauts navigate during **deep space** travel. This is still quite far off in the future.

The largest stars live for "only" a handful of millions of years, while a medium-sized star like our Sun is expected to last at least ten billion years. Neutron stars may last much, much longer because they are already the "corpse" of a giant star that exploded.

Occasionally two neutron stars collide. When they do, it leads to an explosion that has millions of times the energy that the Sun gives off. The explosion can also release **gravitational waves**, or ripples in space.

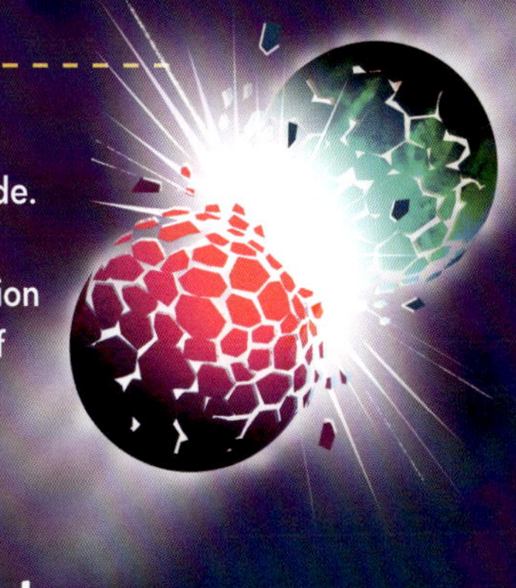

DIFFERENT KINDS OF LIGHT

There are all types of light around us here on Earth—some that occur naturally and sources that are made by humans. A lot of light is harmless, but in some cases, too much of the wrong kind of light can do major damage to living cells. For example, long exposure to X-rays or a blast of gamma-ray radiation from a nuclear bomb is dangerous. Objects in space also give off all types of light. Scientists need a variety of telescopes in different locations—including in space!—to detect all of it.

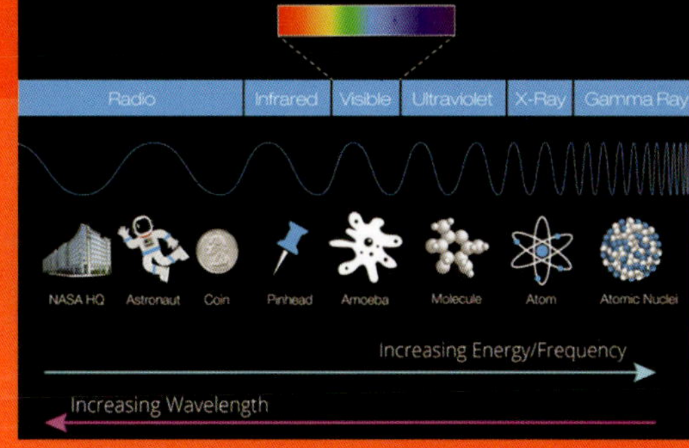

Light can take on different forms, as shown in this illustration. This includes lower-energy light like radio waves, microwaves, and infrared light. Then there is visible light, the kind that human eyes have evolved to detect. And finally, there are higher-energy lights, such as ultraviolet, X-ray, and this might be surprising, but a piano keyboard can be a good metaphor for light. On one end (with the deep notes), there is light with longer wavelengths called radio waves. As you move to the right, there are other kinds of light with shorter wavelengths, including microwaves, infrared, visible light, ultraviolet, X-rays, and gamma rays. Only visible light can be detected by human eyes.

TrES-2b:
The Planet Cloaked in Darkness Forever

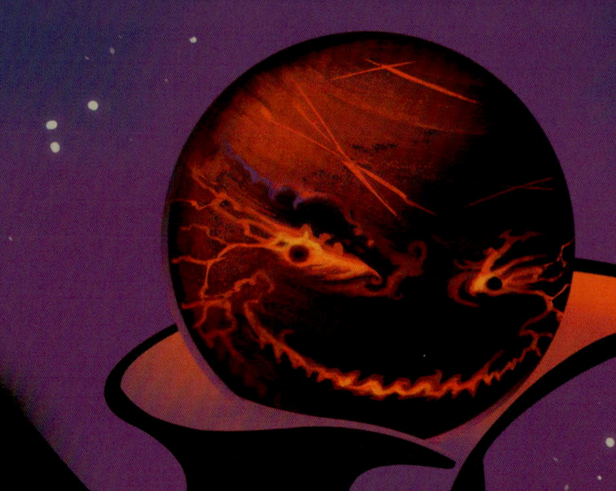

IF YOU ARE AFRAID OF THE DARK, this planet is definitely *not* for you. TrES-2b is the darkest planet we have ever discovered. It is beyond the turn-off-your-lights-in-your-bedroom-but-you-can-still-see-under-the-doorway dark. This planet is more I-can't-see-my-hand-in-front-of-me, haunted-house kind of dark.

TrES-2b absorbs 99 percent of the light from its parent star, meaning that it reflects barely any light. It's darker than charcoal, blacker than a lump of coal, and beyond the black of black acrylic paint.

Most things around us do not give off their own light, or at least the kind of light we can see with our eyes. This is a little weird to think about, but it's true. An apple isn't red because it's glowing red. We see it as red because an external light source—sunlight or light from a light bulb—reflects most of the red part of the rainbow back, while the rest, the green and blue light, is absorbed by the apple. And a field of green grass looks green because the blades of grass absorb most other colors and reflect the wavelengths of green outward.

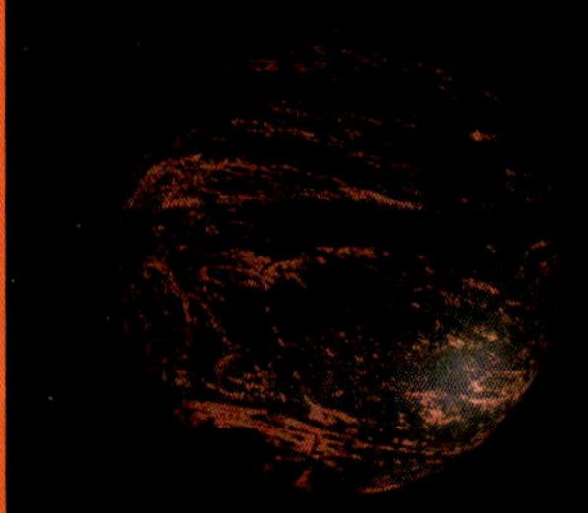

Exoplanet TrES-2b is a place of forever night, as an artist has drawn in this illustration. Think of it as a very, very large lump of coal nestled up closer to its planet star than Mercury is to our Sun.

Why do objects or people reflect some colors and not others? It is because of their atomic and molecular structure. In other words, some atoms and molecules are built to absorb, or keep, certain wavelengths and reflect, or reject, others.

This brings us back to the wild dark world of TrES-2b. Scientists think that it is so dark because the planet's atmosphere is made up of chemicals that absorb light extremely well. The kinds of chemicals that could do this, such as vaporized sodium and potassium as well as titanium oxide in gas form, are uncommon here on Earth and, so far, unknown on exoplanets.

Astronomers use the term hot Jupiter to classify massive planets (like Jupiter) that have a very short, fast orbit because they are much closer to their star than Jupiter is to the Sun. This artist's illustration shows a dark hot Jupiter like TrES-2b close to its own parent sun. Hot Jupiters are weird, exotic planets unlike anything else in our Solar System. But because of their size and closeness to their suns, they are easier to find than other smaller, dimmer exoplanets.

The idea of an entire planet in eternal darkness is frightening enough. But more than that, TrES-2b is a **hot Jupiter**, the type of exoplanet that is at least the size of Jupiter or bigger, but closer in its orbit to its star than Mercury is to our Sun. (Astronomers know of hundreds of hot Jupiters, but those are just the ones they've found so far. There are undoubtedly thousands or even millions in the Milky Way!) This means TrES-2b is being blasted by its star's light like the glare of a zillion spotlights.

All the light and heat from the star goes somewhere—and that's straight into the planet itself. TrES-2b holds on to that light and radiates it like a giant burner on an electric stove. The surface temperature on TrES-2b is about 1,800°F. So, if you're on this planet, don't touch anything.

In fact, astronomers first discovered TrES-2b by using a couple of telescopes in Arizona and California. While these are professional telescopes, they are quite small, with mirrors only ten centimeters, or less than four inches, across. Astronomers then turned to a much bigger telescope on

Hawaii to follow up on the **transit** that these smaller telescopes had picked up. Finally, scientists turned to the Kepler telescope, which was operating in space, to get an even better look at this eternally dark world.

Astronomers have discovered many exoplanets by using the transit method. A transit is when a planet passes between a star and the telescope looking at it. We can see transits in our own Solar System. For example, we can see a transit from Earth when Venus travels between us and our Sun.

Transits are a useful tool in astronomy because the planet passing in front of its star will very slightly dim the light we capture of the star. That dimming of light is shown in a special graph called a light curve, which illustrates the light captured during a certain period of time. So when an exoplanet passes in front of its parent star, the light curve shows the dip in the captured brightness.

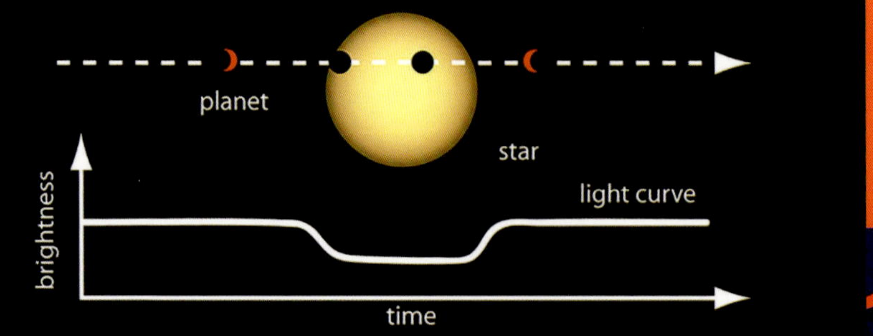

One of the first exoplanets ever discovered was a hot Jupiter named 51 Pegasi b in 1995. While there were hints that exoplanets existed before, this one kicked off the field of exoplanet research. In 2019, two scientists were given the Nobel Prize for discovering 51 Pegasi b.

With a size about 1.5 times that of Jupiter, TrES-2b orbits its parent star at a fraction of the Earth's distance to the Sun, making one complete trip around in just 2.5 days!

The Kepler telescope was a NASA mission that operated in space beginning in 2009 and looked for transits. Even after it lost two of its gyroscopes (wheels that help point the spacecraft), engineers figured out how it could still collect data. By the time the spacecraft ceased operations in 2018, it had observed half a million stars looking for planets.

Astronomers use different kinds of telescopes on the ground and in space to study the Universe. Since not all light can penetrate our Earth's protective atmosphere, we send telescopes into space to study some kinds of light and build very large telescopes on the ground for others.

TELESCOPES: GROUND VS. SPACE

The more you learn about astronomy, the more you find out that it often takes a lot of people and a lot of telescopes to make a discovery. While the scientists are almost always on the ground, sometimes telescopes are on the ground, and sometimes they are in space. Why put some telescopes in space?

Objects in space—including our Sun—give off all different kinds of light. Some of these kinds of light, such as X-rays and gamma rays, would kill us from overexposure. Luckily for us, our atmosphere blocks this kind of harmful radiation. However, if we want to study what the Universe is doing in these kinds of light, we need to launch telescopes above Earth's atmosphere.

Second, the Earth's atmosphere is constantly moving as air and weather flows around the planet. Even when it's a clear night with no obvious cloud cover, the atmosphere can blur the observations of faint and distant objects in space. By sending telescopes out into space, we can eliminate that issue.

Then why don't we send all of our telescopes into space? One reason is it's a lot more expensive and challenging to launch things into orbit. Even though we've made a lot of progress on rockets and spaceflight, there are limits on the size and weight of things we can launch. So telescopes here on the ground give us a chance to build much larger, more sophisticated facilities to learn even more about the Universe we live in.

HD 189733B:
THE IT'S-RAINING-GLASS-SIDEWAYS PLANET

MOST OF US COMPLAIN ABOUT the weather from time to time. You have to walk to school and it's raining or snowing? That's not fun. Too hot to be outside for more than a few minutes? Oh no, get ready to melt. But astronomers have found that there are much, much worse places weather-wise in space. One alien planet with particularly horrific weather is called HD 189733b. Don't let the boring name fool you. This killer world probably should be known as the "Worst-Weather-Ever Planet."

Let's go through the ways that weather on this exoplanet, a planet outside of our Solar System, is completely wretched. Start with the wind. It howls around the planet at speeds of up to 5,400 miles per hour. For some perspective, consider that the record for wind speeds during a hurricane on Earth is about two hundred miles per hour, and the highest wind ever recorded anywhere on our planet was two hundred and thirty miles per hour. If we ever had winds of one thousand miles per hour, they would tear buildings apart and

This artist's illustration shows a beautiful exoplanet, almost Earth-like in its appearance. But don't be fooled, as looks can be deceiving! HD 189733b is a huge gas giant, like Jupiter, but very close to its parent star. This exoplanet's weather is notoriously bad, with howling winds, raining molten glass (sideways!), and a scorching temperature to make it even more terrible!

shred trees. Imagine dealing with gusts that are five times as fast, or seven times the speed of sound, and that's how fast the wind on HD 189733b is blowing.

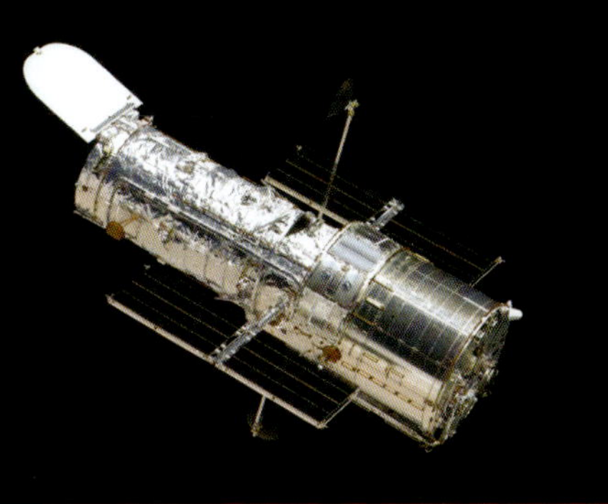

The Hubble Space Telescope, shown in this photo, launched in 1990, before the first exoplanet was discovered. Since technology has advanced, Hubble has studied exoplanets and even directly detected an exoplanet's atmosphere.

And HD 189733b has much more in store for would-be weather chasers. This planet is, in fact, best known for its very evil habit of raining glass sideways. There's an old expression (and a Taylor Swift song) that talks about a death by a thousand cuts. This would be more excruciating, like death by a billion cuts.

This disastrous, razor-sharp precipitation comes from the atmosphere on HD 189733b, where its clouds are made up of particles of **silicates**, a saltlike substance that contains the elements silicon and oxygen. On Earth, we make glass by heating silicates. On HD 189733b, the planet makes glass as the heat from the nearby star turns the clouds of silicates into molten rain.

This blowing-glass-furnace type of atmosphere is what gives HD 189733b a deceptively inviting blue color from a distance, similar to the "blue marble" look that astronauts see of Earth from space. However, HD 189733b's color comes from the star it orbits as the planet reflects more of the blue part of the rainbow off its atmosphere's dangerous particles. It definitely does not reveal a balmy tropical ocean to dip your toes into, like on Earth.

Scientists using the Hubble Space Telescope were able to uncover even more information about this slasher world by using an amazing astronomical detective technique. When the alignment is right, the planet passes in front of the star from our vantage point on Earth. Even though the planet is much smaller than the star, scientists can see the tiny changes in light from the star when this happens. Astronomers call this a transit, and it can tell them a lot about a planet like HD 189733b, including that it has more methane in its atmosphere than they expected. Great. Now there's an unbreathable gas to worry about too on this planet!

So we have wind that will pass through anything, rain that will cut you to shreds, and an atmosphere you can't breathe. It's time to imagine what happens when those things take place at the same time. The glass rain whips around the planet, eviscerating anything or anyone in its path. The only happy news is that astronomers don't think there's any surface to stand on—it's a gas giant planet like Jupiter—so there probably aren't any living things to get pummeled by this lethal weather.

This artist's illustration shows a cornucopia of various exoplanets discovered at different times over the past decade. From hot Jupiters and lava worlds, to stinky planets and dark nightmarish gas giants, it's a weird Universe out there.

Another thing HD 189733b has in common with Jupiter is that it's about the same size, but it orbits its star more closely than our nearest planet (Mercury) orbits our Sun. Instead of being cold like Jupiter, HD 189733b sizzles at a temperature around 2,000°F. This might give this planet the trifecta of bad weather leaving any visitor blown, shredded, and baked.

Many types of planets have been identified outside of our Solar System. Some of them resemble planets like Jupiter, Neptune, or even Earth, but others are very different in terms of size, location, and other characteristics.

Astronomers have found thousands of exoplanets across the Milky Way galaxy since the first confirmed discoveries in the 1990s. Telescopes both on the ground and in space are regularly discovering new exoplanets.

Scientists study the habitability of exoplanets, which means whether life forms like humans could survive and thrive there. The habitable zone is a region around a star where a planet could orbit and theoretically contain liquid water. HD 189733b is so close to its star that it's too hot to have liquid water, meaning astronomers would not consider it to be in a habitable zone.

A colorful collection of twenty-five different hot Jupiters were all studied with the Hubble Space Telescope. Hot Jupiters can have many differences in size and makeup, as shown in this artist's illustration.

HOT JUPITERS

HD 189733b belongs to the hot Jupiters grouping of exoplanets like 51 Pegasi b, one of the first exoplanets discovered. In the early days of finding exoplanets (the 1990s), most of the planets discovered were hot Jupiters. At first, some people wondered if this meant there were more of these kinds of planets than smaller rocky planets like Earth.

It turns out, astronomers were finding hot Jupiters first because of the technique being used. By measuring tiny wobbles in the star, astronomers could figure out if there was a planet tugging on the star. The bigger and closer the planet is to a star like hot Jupiters, the easier this kind of signal could be picked up.

As new techniques were developed and new instruments were built and used, many different types of exoplanets began being discovered. One of the ultimate goals is to find a planet like Earth that's in a habitable zone, has an atmosphere, and is also roughly the same size as our planet. This would be the best spot to look for signs of life.

55 CANCRI E: A SCORCHING PLANET ON A SPIRAL OF DOOM

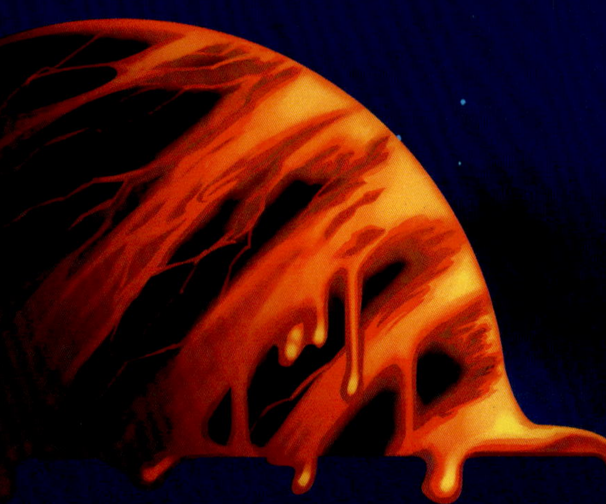

THERE IS LAVA ON EARTH, and there are also huge pools of lava on planets like Venus and Jupiter's moon Titan. However, there is a planet we have not discussed yet, where the entire surface is covered in lava.

Welcome to 55 Cancri e, which is in orbit around a star similar to our Sun just forty-one light-years away from Earth. This exoplanet, a planet outside our Solar System, is roughly comparable to Earth in its size, only about twice as wide as our home planet, and weighs eight times as much. (While that might sound like a big difference, remember that Jupiter is eleven times wider than Earth and weighs over three hundred times as much.) Astronomers refer to exoplanets that are larger than Earth and smaller than Neptune as **super Earths**, but there is nothing super about a visit to this planet.

One of the planet's most horrific characteristics is its location. 55 Cancri e is just 1.5 million miles from its parent star—or sixty-five times closer than the Earth is to the Sun. At this extremely close distance, it whips around its star,

An artist's impression of the lava world called 55 Cancri e shows the planet as it snuggles up to its parent star. Unfortunately for the planet, this kind of snuggling means impending doom as it will eventually be devoured by the star. Until then, conditions on this super Earth will remain miserable, with only oceans of lava to stand on and inhospitable skies to breathe in.

which astronomers have named Copernicus, taking less than three (Earth) days to complete one orbit. This is in contrast to the 365 days it takes the Earth to make the same circuit around our Sun, marking the start of one year to the next.

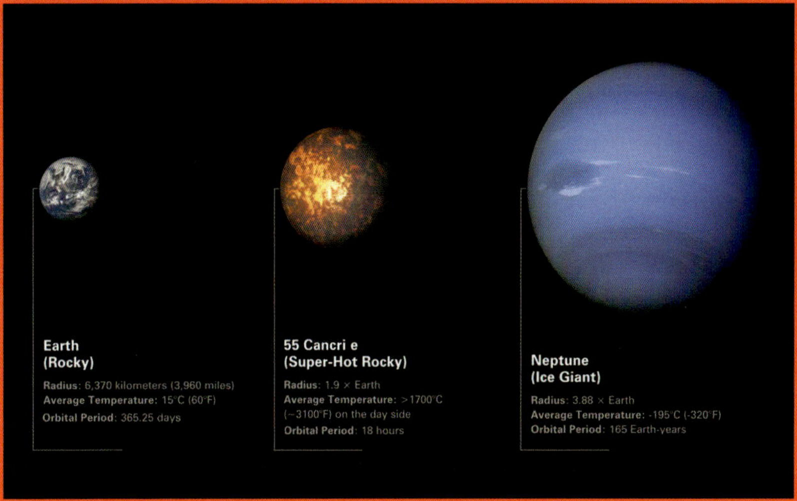

Compare the sizes, temperatures, and orbits of our home planet Earth, the exotic lava world 55 Cancri e, and the ice giant Neptune.

Like other worlds at point-blank range from their star, 55 Cancri e bears the full brunt of what Copernicus dishes out. Soon after astronomers discovered the planet in 2004, they calculated that the side facing its star roasts at temperatures of over 3,000°F, or hot enough to melt iron. Newer data suggests that the temperature on the side facing its star could even reach temperatures of over 6,000°F.

What might the exoplanet 55 Cancri e look like? We can't see this system directly, and we certainly can't travel there with our existing technologies. But this artist's impression shows what scientists think a boiling-hot lava world might be like as it is pulled towards its parent star.

And the air temperatures are just the beginning. Scientists have recently found that 55 Cancri e is covered in oceans of lava. The scene on 55 Cancri e is such an inferno that some describe it like Mustafar, the planet where Anakin Skywalker battles it out with Obi Wan Kenobi in *Star Wars Episode III: Revenge of the Sith.* Once again the reality of space might be much scarier than what science fiction has conjured. While Mustafar had hard surfaces to stand, wield a lightsaber, and land a spaceship on, scientists hypothesize 55 Cancri e is completely covered in molten lava!

Scientists have had to work hard to come up with an explanation for how a planet like 55 Cancri e formed so close to its star. The struggle comes from the fact that, according to what we know about how stars and planets form, 55 Cancri e should never have withstood the chaotic environment near Copernicus as it pulled itself together.

By using newer and more sensitive instruments, and collecting more data, astronomers began unlocking answers. Evidence from the Lowell Discovery Telescope in Arizona released in 2022 showed that 55 Cancri e used to be much farther out from its star but has been spiraling closer and closer over time, as the star's gravitational pull draws it in like a slow-moving tractor beam. This movement has been happening for millions of years, making the sweltering planet hotter and hotter over time.

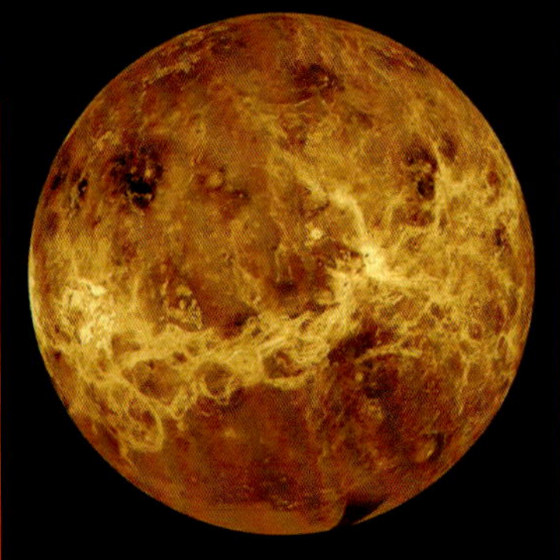
Scientists hypothesize that 55 Cancri e is not the only planet in this system. There is evidence for at least four other planets farther out in orbit around the parent star, Copernicus.

Scientists are still trying to figure out what other deadly tricks 55 Cancri e has up its sleeve. Initially after it was discovered in 2012, scientists thought that the planet's atmosphere would be filled with graphite and carbon—the stuff of diamonds. But in more recent years, data from other telescopes, including the Hubble Space Telescope and James Webb Space Telescope, have cast doubt on that idea. It's possible the skies of 55 Cancri e are filled with gases in a really weird liquid-like state, which would immediately lead to asphyxiation.

In short, 55 Cancri e will scorch you from below and suffocate you from above. We'll take a hard pass on this one.

Astronomers think that 55 Cancri e is not the only planet in this system. There is evidence of at least four other planets farther out in orbit around Copernicus.

55 Cancri e is also known by the name of Janssen, after a Dutch inventor who lived in the fourteenth and fifteenth centuries who is often credited with inventing the microscope and maybe an early telescope. It was given this name through a public voting process orchestrated by the International Astronomical Union.

Super Earths are a class of exoplanets that are more massive than the Earth but smaller than the icy giants like Neptune and Uranus. Our Solar System does not have any super Earths, but scientists think they may be common elsewhere in the Milky Way galaxy.

WHEN A WOBBLE GIVES YOU A WORLD

It can be hard to wrap your mind around how difficult it is to spot—never mind study—exoplanets. Not only are these planets outside our Solar System trillions of miles away; they are also much, much dimmer than the star they orbit. Because they are washed out in the star's light, it took decades before astronomers figured out a way to find them.

In the 1990s, teams of astronomers used an ingenious, though very difficult, technique to discover the first exoplanets. Even though any planet around a star is much less massive than the star itself, it still has a gravitational pull on that star. This, in turn, causes the star to wobble ever so slightly. If you get detailed data of the star's light, you might be able to tease out these subtle changes caused by planets in orbit around them. This technique, called **radial velocity** measurement, turned out to be wildly successful. It resulted in the discovery of hundreds of exoplanets and the 2019 Nobel Prize in Physics being given to astronomers Michel Mayor and Didier Queloz. Today, astronomers have other ways to detect and study exoplanets, but radial velocity remains an important tool in their tool kit.

WASP-12B:

A WORLD BEING EATEN BY ITS OWN STAR

THE SUN, OUR NEAREST STAR, is our planet's source of warmth and light. Not every planet has this kind of supportive relationship with its star. WASP-12b, for example, is an exoplanet under attack by the star it orbits. While many exoplanets are being bombarded by their parent star, WASP-12b is under a full-fledged assault.

This onslaught comes for several reasons. The star that WASP-12b orbits is an enormous beast, and WASP-12b is too close for comfort like other hot Jupiters. It's so close, in fact, that it takes WASP-12b a little over one Earth day to circle this star, turning the planet into a scorching sizzle-fest.

Because of its colossal size, the parent star also has extra pull on WASP-12b. These gravitational forces from the star stretch the planet's atmosphere to the point where it's more egg-shaped than a sphere, which is what they typically are. This may sound like no big deal, but it is.

This artist's illustration of the exoplanet WASP-12b shows the hottest known planet in our Milky Way. This very large and hot world is being cannibalized by its own parent star. WASP-12b is very close—too close!—to its parent star. The proximity of the star and planet means that WASP-12b is almost 4,000°F and stretched into an egg shape by the gravitational force

The gravitational tug from the Moon is what gives us tides on Earth as the oceans move back and forth with the rhythm of the Moon's orbit. The tidal forces that WASP-12b's star

puts on this planet are much, much more extreme. The forces aren't only pulling on the surface; they're pulling on the entire planet, causing it to come apart.

Our Earth and Moon system was captured by NASA's Galileo mission as it flew by in 1996. The Moon and Earth both pull on each other. On Earth, the Moon's pull causes the oceans to bulge out and create tides.

As the atmosphere swells and becomes distorted, it's simultaneously getting pulled off the planet each time it circles around its star. Astronomers estimate that billions of tons of gases from the planet's atmosphere are being ripped off the planet every second. Once the gases leave, they're eaten by, or incorporated into, the star. Standing on this planet would be like living in a constant tornado with everything around you—including the sky—being torn away into space.

There's one more piece of bad news for WASP-12b. Its orbit is on a death spiral that brings it slightly closer to its star every pass. So astronomers predict that the star will completely consume the planet in about ten million years. While that sounds like a long time, it's like a blink of an eye in space and in the history of time. To put it another way, the time between the age of the dinosaurs and present-day Earth is much greater than the time until this planet's destruction. Keep in mind that our own Sun has been around for about five *billion* years, and we expect it to last five billion more.

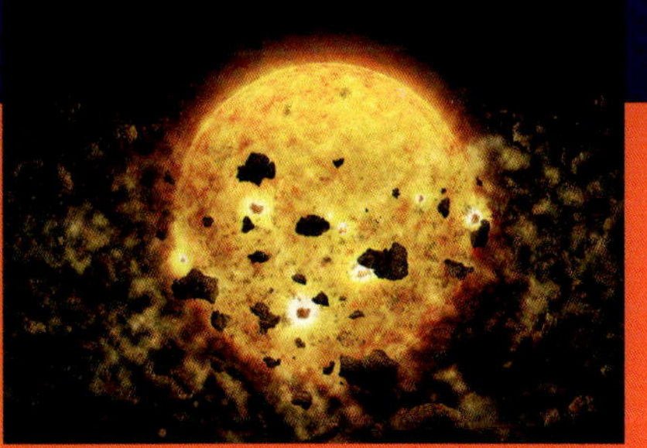

What might a planet look like as it gets closer to being cannibalized by its parent star? This artist's illustration shows the violent destruction of a young planet as it is obliterated.

Not only is this cosmic game of tug-of-war between the star and planet ripping WASP-12b apart; it's also heating the planet from within. Astronomers think the surface temperatures have soared to about 4,000°F, which is hot enough to melt titanium.

Like other exoplanets, this ill-fated world was discovered despite some serious odds against doing so. Given WASP-12b's distance from Earth (of about 1,200 light-years) and its tiny size, its star puts out about three thousand times more light than it does. So to find WASP-12b, astronomers used the transit method as they have with many other exoplanets. Once they knew it was there, additional scientists used the Hubble Space Telescope and other telescopes to make new observations, like how WASP-12b is being eaten by its parent star!

So if you're looking for a spot in the Universe to watch the destruction of a planet, this would be a great destination. Just be warned, however, it could be your last.

To hunt for exoplanets, scientists use different observatories such as NASA's Kepler and Transiting Exoplanet Survey Satellite (TESS) missions in space, or the Next Generation Transit Survey in the Atacama desert. These missions search for small dips in the brightness of hundreds of thousands of stars in the Milky Way. As an exoplanet transits between its parent star and Earth, it briefly dims the light reaching us, which can be captured by telescopes.

The name WASP-12b refers to how this exoplanet was discovered: a project known as the Wide Angle Search for Planets, which uses telescopes around the world to look for transits. This was the first planet found around the twelfth star studied, which is where the 12 in the name comes from. The star itself is known as WASP-12a, so the planet is called WASP-12b. If there was a second planet discovered in this system, it would be called WASP-12c, and so on.

WASP-12b is about 1,200 light-years from Earth in the direction of the constellation of Auriga, the Charioteer.

While the star is wreaking havoc on the planet, WASP-12b is also getting a little revenge. Stars slow down as they age. Research shows that the star's rotation is slowing down in this system because of interactions with WASP-12b, making the star act older than its age.

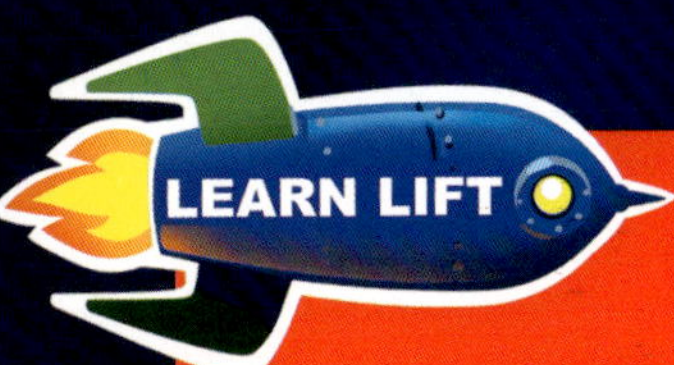

ARE WE ALONE?

One of the biggest questions we have about space has to do with us: Are we alone? The hunt is on for any life within our Solar System, but this would be on the microscopic level—not anything or anyone we could communicate with. The idea of intelligent life outside of Earth (aka aliens) has been the subject of countless books, shows, and movies. But what do scientists think?

As scientists, they need data, and there have been several projects devoted to gathering evidence of whether we are alone. One of the most famous is the Search for Extraterrestrial Intelligence (SETI). This is a name that covers different efforts to use telescopes to pick up signals—intentional or accidental—from civilizations beyond Earth. So far, nothing.

Another way astronomers are looking for signs of life in the Universe is through the study of exoplanets. The thinking goes that if we can find a planet that is similar to Earth in most ways—temperature, atmosphere, age—then we could focus future telescopes to study it to see if there is any proof of life.

So far, we've found nothing. But who knows what we might discover in the future?

ETA CARINAE:
AN EXPANDING, EXPLODING BRAIN IN SPACE

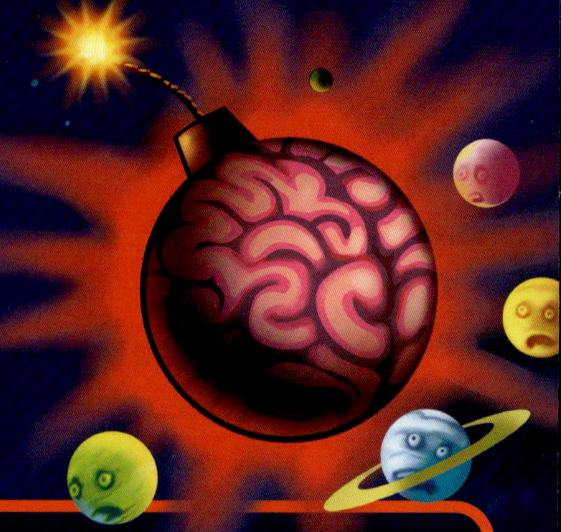

WHAT'S GOING TO BE THE NEXT STAR in our galaxy to explode? Astronomers don't know for sure, but one favorite for a ticking time bomb is the strange star system called Eta Carinae.

Eta Carinae is super weird. In the middle of the 1800s, astronomers noticed that it glowed super brightly. For about twenty years during the nineteenth century, Eta Carinae looked like the brightest star in the sky except for the star Sirius, which is much closer to Earth. Astronomers dubbed this episode the Great Eruption, but they weren't sure what was going on.

Today we know a lot about what was happening with Eta Carinae, and more importantly, what could still happen in the not-so-distant future. Eta Carinae is a loose cannon ready to explode, and a big threat to its stellar neighbors. Fortunately, here on Earth, Eta Carinae won't cause any real damage when it explodes, because it is about 7,500 light-years away. That's about forty quadrillion miles away, which is really far and out of the danger zone.

The Hubble Space Telescope has watched the star system Eta Carinae for over twenty years. The aging massive stars, prone to violent outbursts, previously spewed out material into this brain-shaped bubble in the 1850s. Eta Carinae is in our Milky Way galaxy, about 7,500 light-years from Earth.

To begin with, astronomers have figured out that Eta Carinae is not just one star, but two or possibly even three stars. It's been hard to figure out what is going on with the stars in Eta Carinae, because the stars themselves are entombed in a strange space cocoon that scientists call the Homunculus Nebula.

In the entire vast Universe, a star explodes every second. But in our own much smaller galaxy, we only see one explode maybe every fifty years. NASA's Hubble Space Telescope caught this star (in bright blue in the lower left) after it had exploded in a galaxy about seventy million light-years away from Earth. This supernova explosion briefly outshined the entire galaxy it lives in, with the energy it output as it exploded equal to the brightness of about five billion Suns.

While the name *Homunculus Nebula* sounds like a villain in a superhero story, it's the label for clouds of glowing, superhot gas that formed after being ejected during the Great Eruption, and that look like a brain. So there's an expanding, exploding "brain" in space. Pretty villainous after all.

You might think anything nearby Eta Carinae is in the clear because the Great Eruption happened a century and a half ago. But Eta Carinae isn't done yet.

Eta Carinae is about to go kaboom because one of the stars in the space brain is giant. The very biggest stars are the most unstable and live the shortest lives compared to middle-sized stars like our Sun. The Great Eruption may have been some of the last gasps from this gigantic star before it ultimately blows itself to smithereens. Astronomers think that Eta Carinae could be the next star in our Milky Way galaxy to completely blow itself up as a supernova.

Supernovas are amazingly powerful star explosions that can create enough light to outshine an entire galaxy. (Keep in mind, most galaxies have many hundreds of millions, if not billions or even a trillion, stars!) In addition to the kind of light we can see with our eyes, a supernova unleashes more powerful forms of light—or radiation—in the form of X-rays and gamma rays.

There's a reason your doctor or dentist might still give you a lead apron when you get an X-ray and why gamma rays supposedly turned Bruce Banner into the Hulk. These kinds of more energetic light can warp and kill the cells in your body and cause other bad outcomes. Remember, we're far enough away on Earth, so we're safe. But if you happen to be much closer to Eta Carinae, you might be in a whole lot of trouble. Anything or anyone within twenty-five or fifty light-years or so when Eta Carinae goes supernova will be vulnerable to lethal doses of X-rays, gamma rays, and an overall really bad day.

An illustration of a very bright and energetic supernova explosion. When Eta Carinae eventually explodes, it will (safely) light up our night sky with a bright spot of light, like this illustration, and will likely be visible during the day from Earth.

This ghostly image shows an exploded star called Cassiopeia A (Cas A) with its expanding shell of material that is slamming into the gas shed by the star before it exploded. The orange and pink tufts are from the inner shell of the supernova remnant, along with blobs of oxygen, sulfur, argon, and neon.

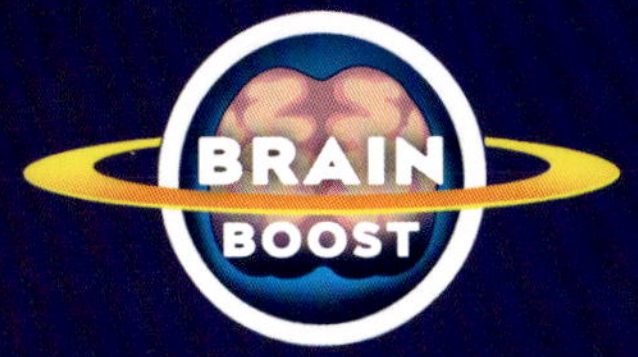

Astronomers think the Homunculus Nebula didn't exist before the Great Eruption, which is what pumped out all of that stuff from the stars. Whenever Eta Carinae does explode as a supernova, the material in the Homunculus Nebula will help seed the area around it and could form many new stars in the future.

The two known stars in Eta Carinae are much bigger than our Sun. The "small" one is about thirty times as massive, and the larger one holds about ninety times more **matter** than the Sun. If you were to drop the larger one into the Solar System, it would reach all the way out to Earth's orbit.

Even though astronomers have been entranced with Eta Carinae for hundreds of years, they didn't know it was actually two (or three) stars until late in the twentieth century, thanks to observations from telescopes like the Hubble Space Telescope.

WE ARE STAR STUFF

We are made up of stars. Literally.

Scientists think right after the Universe was created in the Big Bang, there were mainly only two elements: hydrogen and helium. Since there are lots of other elements now, where did they come from? The answer is from inside stars.

Most of the elements that we need for life—including oxygen, iron, calcium, and so on—were created inside stars. For example, when some ancient stars ran out of fuel, they exploded as supernovas and sent these newly manufactured elements out into space. These elements were swept up into clouds of gas and dust, which eventually formed the next generation of stars and planets. This cycle continued to allow Earth to form and all of us to be here today. So the next time you look up at the stars, you may want to thank them for the life you have because they are our cosmic ancestors.

CRAB NEBULA:

A VORTEX OF DEADLY ENERGY

FEW THINGS MIGHT SOUND FREAKIER than a giant crab in space. Even though this version isn't a giant crustacean, it is just as bizarre. The Crab Nebula, as it is known today, is one of the most investigated objects in the Universe. It was first spotted in the night sky by Chinese, Japanese, and Arabian astronomers in 1054 AD. Other people may have also noticed it, but historians have confirmed historical records from these regions. This "new star" was about four times brighter than Venus and visible during the daytime for almost a month.

About seven hundred years later, an English astronomer named John Bevis looked at the "new star" through the recently developed technology of a telescope and noticed it wasn't a regular star. Instead, it was fuzzy. Another astronomer of the eighteenth century, Charles Messier of France, thought this object was Halley's Comet. Then in the 1840s, an Irish astronomer named William Parsons thought it resembled the pinchers of a crab. Even though other people didn't see that shape, the name Crab Nebula stuck.

The stellar graveyard that is called the Crab Nebula has been studied by astronomers for centuries. This image from the James Webb Space Telescope shows the golden cage-like structure and white ghostly material that are captured in infrared light.

There is no mystery around the identity of what the Crab Nebula is today. The backstory of the Crab is that a really big star ran out of fuel. Once that happened, the star no longer had the strength or pressure to hold up the outer layers of the star, which fell downward due to gravity. Without that pressure, the outer layers crashed down on the core—creating a stellar corpse—and then bounced back out into space in a huge stellar explosion called a supernova.

Inside the wispy ringlike area in the core, at the very center of this image taken by the Hubble Space telescope in visible light, is the stellar corpse. A stellar corpse contains the remains of the original star that becomes a rapidly rotating neutron star.

A supernova explosion can leave behind an ultradense nub of the star called a neutron star, which gets its name because it is essentially only neutrons packed together. Sometimes neutron stars rotate quickly and give off beams of light or radiation as they spin. Astronomers called these **pulsars**. And at the center of the Crab Nebula is a pulsar.

The pulsar in the Crab gives off pulses of light while spinning a spectacular web of energy. This generates powerful beams, or **jets**, that blast out of the top and bottom of the pulsar. These twin beams appear to flash on and off as the pulsar spins around at a remarkable thirty times per second, or over three million miles per hour. These jets help to push the whole nebula outward into space.

Meanwhile, there is a wind screaming away from the pulsar in the middle of the nebula, which itself is made up almost entirely of hydrogen that came from the dying star. The entire system has supercharged magnetic fields that contribute to the jumbled chaos surrounding the pulsar. As particles travel around the pulsar, riding at nearly the speed of light, they follow the magnetic field and release radiation as they go.

So, what John Bevis saw as a fuzzy blotch from his telescope in the 1700s, is technically a swirling vortex of magnetic fields, energy, and particles. This hair-raising view can only be seen when using telescopes that are sensitive to the most powerful kinds of light, such as X-rays. Getting into the Crab would be asking to be sloshed around inside a lethal space blender.

To recap, the Crab Nebula is not just a bright star. It's a treacherous cosmic vortex, crawling out into space and creating creepy, threadlike fingers that pulsate out into interstellar space!

Comets are like dirty snowballs orbiting through our Solar System. Some, when they get close enough to Earth, can put on quite a show in the night. This image shows comet Halley as snapped by William Liller from Easter Island in 1986. Before our telescopes were as powerful as they are now, objects like comets could be confused for exploded stars in the night sky.

Charles Messier, the astronomer who mistook the Crab for a comet, went on to name 110 astronomical objects. Each object was given his name and a number. The Crab is Messier 1 (M1), and that name is still used today.

There are pictographs—one of the earliest forms of writing, which uses pictures to convey data—that look like the Crab Nebula in Chaco County, New Mexico. While scientists have not been able to officially tie the two together, many researchers believe this is a recording of the explosion that created the Crab Nebula.

The Crab Nebula is in our Milky Way galaxy at a distance of about 6,500 light-years from Earth. The pulsar driving the nebula is about ten miles across, while the nebula itself is about sixty trillion miles wide.

HOW HUMANS HELP SCIENCE IN SPACE

The Crab Nebula is one of the best examples of how different parts of mysterious cosmic objects—and the science that creates them—are revealed only by looking at the full range of light. While scientists knew about X-rays and gamma rays by the early twentieth century, they couldn't study them because any telescopes or instruments needed to be above the Earth's atmosphere to do so.

The dawn of the Space Age in the 1960s changed everything. Once NASA and other space agencies figured out how to send machines into space, it opened the door for scientists to join. Over the decades, astronauts in space have helped launch and repair telescopes and instruments. There are astronomical telescopes on the International Space Station today, for example, that have been installed by astronauts working there.

Perhaps the most famous event involving humans working on telescopes in space was the repairs of the Hubble Space Telescope. Shortly after it was launched in 1990, scientists and engineers realized that one of Hubble's mirrors was flawed. Fortunately, Hubble had been designed so that astronauts could return to it and swap out old instruments and broken pieces. This flawed mirror made this urgent.

Three years after its launch, astronauts brought a new camera that compensated for the flawed mirror, plus other new equipment. The astronauts made a series of very tricky repairs—in space!—that corrected the problems the telescope was having and saved the mission. Astronauts returned to Hubble four more times, the last time in May 2009.

A group of astronauts trained intensively on Earth before launching into space in 1993. Their job was to repair the Hubble Space Telescope while it was in orbit, in a series of five days of space walks. In this photo, astronaut Story Musgrave was attached to a Remote Manipulator System arm to be brought to the top of Hubble for repairs. A second astronaut, Jeffrey Hoffman, worked inside the Space Shuttle's payload bay.

SAGITTARIUS A*:
THE BLACK HOLE AT THE CENTER OF OUR GALAXY

THERE'S AN OLD-FASHIONED SAYING: "There's a tempest in a teapot." It refers to a major overreaction, or when there's too much being made out of very little. Whoever thought of this phrase did not know that there happens to be a cosmic tempest in a celestial teapot inside the **constellation** of stars called Sagittarius.

Inside Sagittarius, there's a group of stars that resembles a teapot. If you draw a line with your finger along two of the stars in this teapot shape, you will find the center of our Milky Way galaxy. A **supermassive black hole** exists in that very heart of the Milky Way and is called Sagittarius A* (Sgr A* for short, which is pronounced *saj-ay star*). It's a little confusing; the black hole borrows the name from the constellation, but it's a separate object.

Before going any further, let's take a moment to talk about what a black hole is. Black holes are regions of space where there are huge amounts of mass, or stuff, packed into a relatively small space. This creates superstrong gravity that keeps anything—including light—within a certain distance of the black hole from ever escaping.

This image shows intense activity of the bustling "downtown" region around our supermassive black hole called Sagittarius A* which is in the bright white patch in the middle-right of this image.

The supermassive black hole at the center of our Milky Way galaxy has about four million times the mass of our Sun in a relatively tiny region. Our galactic supermassive black hole is no shrinking violet. It's responsible for sending blasts of energy out into space, ensnaring stars, and dictating how many stars form around it.

This artist's illustration shows a close-up of a supermassive black hole like Sagittarius A*. The black hole is surrounded by hot gas colored in red and yellow, which acts as fuel for the black hole engine. Power generated by the engine flows away from the black hole via jets of high-energy particles.

While that sounds terrifying, it's good to put our galaxy—and its resident giant black hole—into context. The first thing to realize is our distance from Sgr A*. Thankfully, it's over 26,000 light-years from Earth, which means that we're very far away and safe from it.

Next let's talk about how the galaxy is constructed. Our Milky Way is a **spiral galaxy,** which means it looks like a giant pinwheel. Another way to think about a spiral galaxy like ours is to compare it to a city. The Solar System, including Earth, sits in a quiet neighborhood on the outskirts. We live in one of the Milky Way's spiral arms, about two-thirds of the way out to the edge. Not much happens out here, and that's a good thing.

The Event Horizon Telescope (EHT) released the first-ever image of our Milky Way's supermassive black hole's shadow in 2022. Then in 2024, it captured this view showing the gigantic swirling vortex of the magnetic field around the shadow of Sagittarius A*.

The center, where Sgr A* lives, on the other hand, is the galaxy's bustling downtown. Like city centers here on Earth, there is a lot of action and energy. Astronomers have been trying to better understand this important region for a long time. However, there's a lot of dust and gas between us in the outskirts and the downtown of the galactic center. So we couldn't see much of this area with telescopes that detect only visible light.

The past couple of decades have changed that. Our telescopes in space can now see what's happening in the galactic center in more kinds of light, from radio waves to X-rays. In fact, astronomers in 2022 revealed the first image of the black hole in the center of the Milky Way. By virtually linking eight radio telescopes scattered around the globe to create an Earth-sized telescope called the Event Horizon Telescope (EHT).

This gave the scientists the ability to see enough detail to, for the first time, see material right up to the edge of the black hole. Since those first EHT observations, scientists have returned to study Sgr A* again and again. Just recently, they found that Sgr A* is a vortex of magnetic fields. This makes the edge of the black hole a scary doomsday scenario of intense gravity and powerful magnetic fields that can disorient and destroy, like a Bermuda Triangle–type area you want to avoid at all costs.

The reasons listed above hopefully give enough reason to stay away from Sgr A*, but there are more. The heart of our Milky Way is a crowded place. There are millions of stars whipping around Sgr A*, with many of their paths being dictated by the gravitational force from the black hole. If you went to this part of the galaxy, you would see a much more crowded night sky with stars seemingly covering the darkness of space.

Some of the stars there are very massive, and they have fierce winds that blow off of their surfaces, battering the surrounding areas. These winds send material from their outer layers toward Sgr A*. If these stars are within a few light-years of the black hole, Sgr A* will ingest that stellar material, giving it fuel to add a little more mass and grow bit by bit.

On top of these hazards, the whole region is lit up brighter than any collection of neon lights and billboards you might find in a city center here on Earth. And much of that light is the high-energy, super-dangerous kind: mostly X-rays. There are outbursts of ultraviolet light and gamma rays too. Exposure to this kind of light, or radiation, would be fatal for life as we know it. So the downtown of our cosmic metropolis is an exciting place, but way too dangerous to venture into.

Our Solar System, containing the Sun and eight planets, is about two-thirds of the way out from the center of the galactic pancake we live in, more formally known as the Milky Way galaxy. Our Solar System travels in an orbit around the center of the galaxy at a speed of nearly 200 miles per second, completing one orbit around the center of the Milky Way about every 230 million years!

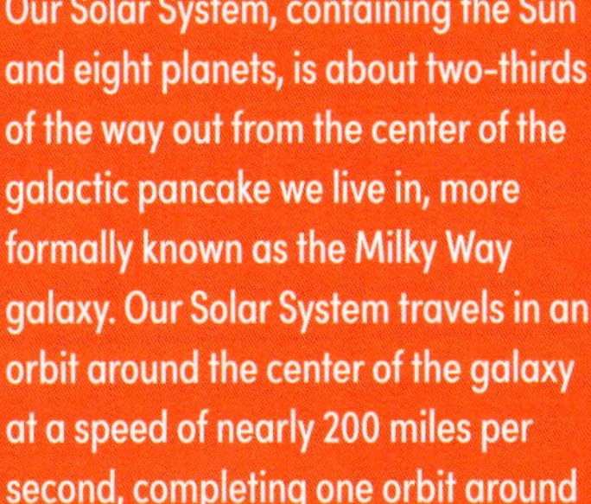

Constellations are patterns of stars officially recognized by the International Astronomical Union, the global organization in charge of naming things in space. There are eighty-eight official constellations.

A well-known group of stars that is not officially a constellation is known as an **asterism**. The teapot in Sagittarius is an asterism, and so is the Big Dipper in the official constellation of Ursa Major.

The discovery of Sgr A* is credited to two astronomers, Bruce Balick and Robert L. Brown, who found a very bright radio signal coming from the center of the galaxy in 1974. It wasn't until 1982 that astronomers gave it the name of Sgr A*.

OUR GALACTIC PANCAKE

It might be a little strange to think of our home in space as a breakfast item, but imagining our Milky Way like a galactic pancake can be very helpful. The spiral arms where Earth is lie in the pancake, which is incredibly wide and flat. The Milky Way is about one hundred thousand light-years across, but only about two thousand light-years thick in most places. The exception to this is in the middle of the galaxy, where Sgr A* sits. This central region is like having a baseball dropped into the middle of the pancake.

Where did our Milky Way come from? Astronomers think it started to form about twelve billion years ago through a series of mergers and collisions of smaller galaxies. Keeping with the comparison to food, it's like smaller bits of pancake batter got mixed together and then were eventually "baked" into the galaxy we live in today. Want one more food-related connection? The word galaxy comes from a Greek word meaning *milky circle* or *milky way*. Who's hungry now?

WHAT LIES BEYOND

Have you had enough yet with Io, HD 189733, and the Crab Nebula? Well, go beyond the edges of the Milky Way, and there are some even creepier sites to explore.

The Milky Way galaxy is but one of billions of galaxies in the Universe. Some of these galaxies are home to even more horrid objects that are behaving worse than you can imagine. Black holes are sprinkled across the Universe like deathly potholes you can't see. There are magnets millions of times more powerful than anything we can make on Earth that will tear you apart atom by atom. Ouch!

Galaxies themselves smash into one another, ripping each other apart and disrupting stars and planets. No need to hang on quite yet, but this will happen to our own galaxy when it collides with the Andromeda galaxy...about five billion years from now. Astronomers have even seen a galaxy that is blasting a gigantic death ray at a nearby companion, causing all sorts of damage to its galactic neighbor.

Galaxies usually do not go through space alone. They are tied to other galaxies through gravity and form packs, or groups, of dozens of galaxies. Our Milky Way belongs to one of these.

Galaxy clusters are the true giants of the Universe. These megastructures have hundreds—and even thousands—of galaxies moving through space while growing and evolving. These galaxy clusters also ram into each other in violent collisions that release stupendous amounts of energy, altering the fate of the galaxies, stars, and planets.

Maybe the scariest parts of the Universe are what we cannot not even see. The two biggest mysteries in space are called **dark matter** and **dark energy**. We don't know what dark matter and dark energy are, but we know what they do. Dark matter is invisible stuff that controls how everything else—planets, stars, galaxies—move. Dark energy is a mysterious force or energy that is pushing the Universe itself apart, and one day, could cause it to rip apart.

Continue reading if you dare!

COLLIDING GALAXIES:
CLASH OF THE TITANS

MAYBE YOU KNOW THAT GALAXIES are made of planets, stars, and other objects. Did you also know that galaxies move across space? This means that sometimes they collide and eat other up! These are massive clashes that take millions of years to happen and change the futures of billions of stars and planets. Our home galaxy, the Milky Way, is actually on a collision course right now with the Andromeda galaxy—but thankfully not for another five billion years.

Earth is in a galaxy called the Milky Way, which is home to about three hundred billion stars. The Milky Way is a kind of galaxy that astronomers have dubbed a spiral galaxy due to gigantic arms reaching out from the center like a pinwheel. At the centers of nearly all galaxies, including the Milky Way, is a supermassive black hole with millions or even billions of times the mass of the Sun.

Galaxies are generally by themselves in space, but sometimes that changes. Stars move around inside a galaxy, and galaxies move across space. Galaxies move both because the Universe itself is expanding, and because they get pulled by the gravity from collections of other galaxies, known as groups and clusters.

Our nearest neighbor galaxy, the Andromeda galaxy, or M31, is about 2.5 million light-years away from us. Andromeda is about 2.5 times the width of the Milky Way, measuring some 260,000 light-years across. Our own Milky Way is destined to smash into Andromeda in about five billion years.

We live inside the Milky Way galaxy, so we can't see our own galaxy from above. We don't yet have telescopes or spacecraft that can fly out beyond our galaxy, since we've only recently exited our own Solar System with NASA's Voyager mission. But we can gaze up at night—from a dark location—to see one arm of the Milky Way. We can also see toward the center of our Milky Way, although vast dark clouds hide the core from human eyes.

If a galaxy gets nudged a little bit by a passing galaxy, it can begin its collision course with another. This is a slow-motion galactic train wreck. Over millions of years, galaxies can pass each other more than once, each time getting closer and closer. When they finally do collide, the results can be catastrophic. Imagine you have two handfuls of marbles that get mixed up on a table with a hole in the middle. If you throw the marbles around hard enough, some of the marbles might be flung off the table and get lost. Some of the other marbles might fall into the hole in the middle. Something similar happens to stars when galaxies collide: Some stars get ejected from the galaxies, while others fall into one of the giant black holes.

The effects of a galactic collision don't stop there. The strong pull from each galaxy will change the other nearby galaxies. If the galaxies have spiral arms, they will become long and distorted. The force from the collision will send waves of energy rumbling through the newly wrecked galaxies. These waves compress the giant clouds of gas and dust, igniting a new generation of baby stars popping off throughout the galaxies.

The giant black holes that live in the centers of the galaxies also get into the game. The black holes themselves collide and merge, eventually turning into one humongous voracious black hole. Before that happens, however, the orbiting black holes send out waves of energy into space once they are within about a light-year of each other. Known as gravitational waves, this energy warps the very fabric of space itself.

So we know a galaxy collision of our own is coming, but here's the good news: Chances are when our Milky Way crashes into Andromeda in about five billion years, many of the individual stars won't collide because there's so much space in between them. (There's actually a lot of space in space! Take our Sun, for example, and the nearest star to us, Proxima Centauri. It's our closest neighboring star, and it's about 4.2 light-years, or about twenty-five trillion miles, away.)

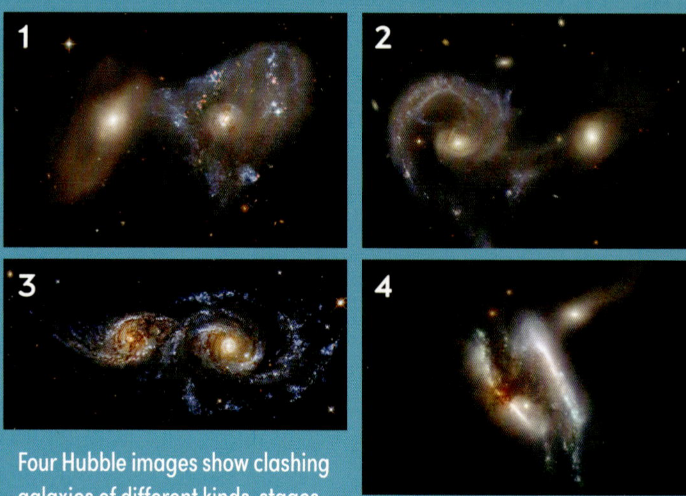

Four Hubble images show clashing galaxies of different kinds, stages, and formations: 1) A burst of stars forming in the galaxy (right) as it merges with a companion (left). 2) A large bright galaxy connected by a bridge of stars, gas, and dust, merging with a smaller one. 3) Three galaxies merging. 4) A pair of colliding spiral galaxies.

Not only are stars spread out, but each star is also tiny in comparison to the distance between them. For example, if the Sun and Proxima Centauri were both the size of peas, they would be separated by two hundred miles—that's the distance between New York City and Boston. Imagine the likelihood of two peas in those two cities rolling right toward each other and running into each other. Most other stars are much, much farther away from the Sun and each other.

Still, when our Milky Way collides with Andromeda years from now, it will go through an extreme galactic makeover on many levels that would be astonishing and strange to witness. At least we won't be around to see it!

What might our night sky look like in three billion years (give or take)? An artist's illustration shows what scientists think the start of the merger between our Milky Way and Andromeda will look like as Andromeda comes in on the left and pulls on our own galaxy.

THE EXPANDING UNIVERSE AND TIME MACHINES

In some ways, the Universe is like an expanding bubble. After the Big Bang, which was a tiny point, everything started moving outward. Over the last 13.8 billion years, everything has been moving away from each other.

A fun way to imagine this is with raisin bread. When the dough is raw, the raisins in the bread are pretty close together. As the dough bakes, each individual raisin moves away from the others. The early Universe in this analogy is the raw dough, and the raisins are galaxies. You can see when the Universe was younger (that is, unbaked), galaxies were closer together than they became as the Universe has expanded (baked). This means that when the Universe was younger, or closer to the time of the Big Bang, everything was more crowded.

Telescopes essentially act as time machines for us. We are seeing things in the Universe as they *were*, and not as they *are* because it takes light so very long to reach us from distant places. We have powerful telescopes that let us see back to very early in the Universe, so we get glimpses of how the cosmos has changed over billions of years.

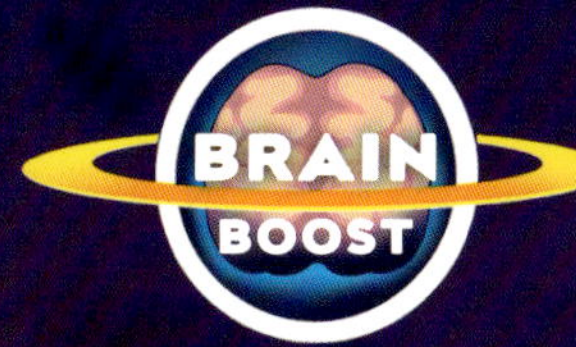

Astronomers estimate that there are more than a hundred billion galaxies in the Universe, each of which has billions of stars. Multiply those two numbers together, and our Universe has a lot of stars in it.

The three main categories of galaxies are spirals, ellipticals, and irregulars, each of which is named after their shape. Some irregular galaxies might be in the middle of a galaxy collision that hasn't finished yet.

Even though most galaxies are alone in the Universe, like islands unto themselves, they often have a larger extended family. They either belong to a group (a few to tens of galaxies) or a cluster (hundreds or thousands of galaxies). Our Milky Way is part of the Local Group that has about fifty galaxies all connected through gravity.

Ellipticals, spirals, and irregulars, oh my! Compare illustrations of the three main types of galaxies in the top row of this graphic with images of galaxies from the Hubble Telescope in the bottom row of this graphic. The Milky Way is a spiral and nearly 50 percent of the other galaxies we can see nearby are also spirals.

MAGNETARS:
DEAD STARS THAT WILL TEAR YOU APART

HAVE YOU EVER USED MAGNETS to drag metal shavings around, pull a toy train along tracks, or hang some artwork from your fridge? Magnets in space may blow your mind. There are objects in space that are such powerful magnets, they can pull your atoms apart until you dissolve. Yes, you read that right!

Magnetars are a type of neutron star, a kind of dead star that is created when a humongous star several times the size of our Sun runs out of atoms to smash together for fuel. The gravity is so strong that the star collapses onto itself and leaves behind a superdense core.

This is like taking your sleeping bag and packing it into its carrier bag—and then continuing to pack it tighter and smaller until the sleeping bag is microscopic. This might seem impossible, but it's a way to describe what happens with these cosmic objects. When neutron stars form, electrons get smashed into protons, turning them into neutrons. Because of

A magnetar, as drawn in this artist's illustration, is a weird object in space. It's a spinning neutron star—the dead core of a massive star that exploded—with a really intense magnetic field. Astronomers discovered magnetars using X-ray light because they have very sudden outbursts of high energy that X-ray telescopes pick up.

An artist's illustration shows a magnetar being born with the magnetic field lines drawn on.

the way they are built, neutrons can pack together really tightly.

The result is a neutron star, or a ball about ten miles across (about the length of Manhattan, New York) with the density of a nucleus of a single atom. If that doesn't sound like it makes sense, imagine this: One teaspoon of a neutron star would have the same mass as an entire mountain here on Earth.

For reasons astronomers don't fully understand, some neutron stars are born with another superpower: supercharged magnetic fields. These are magnetars. All stars have magnetic fields—including our Sun—but the ones in magnetars are millions to billions of times stronger.

These mega space magnets are rare. Astronomers have only found about thirty magnetars in our Milky Way, compared to the billions upon billions of objects in our galaxy. Even so, you should still beware of these scary magnetic monsters.

Think of Magneto, the comic book character in X-Men who can control anything metal using magnetic fields. Magnetars, with the strongest magnetic fields in the Universe, can do a whole lot worse than Magneto.

If a magnetar appeared about sixty thousand miles from Earth, its magnetic field would wipe out every credit card and hard drive on the planet. At only one thousand miles from Earth, the magnetic field from a magnetar would warp all the atoms in your body.

What's the big deal? You might be thinking, *I'm not made of metal.* While that's true, your nervous system uses electrical charges that can be messed up by a strong enough magnetic field. A superpowerful magnet like you'd find in a magnetar would make all the positive charges in your body's atoms go one way while the negative charges pull the other. The atoms in your body would come apart, and you would basically break down piece by piece.

Take heart, however! Astronomers have done a complete scan of the area around Earth and the Solar System, and there are no magnetars anywhere in our vicinity. If you choose to fly out to a magnetar elsewhere in the Milky Way or well beyond in a distant galaxy, however, consider yourself warned.

The bright flare made by the magnetar SGR 1806–20, as seen in this artist's illustration, traveled for fifty thousand light-years before it collided with Earth's atmosphere in December 2004. The burst of energy was so powerful that it disrupted part of Earth's upper atmosphere, causing problems with radio communications. It was the brightest outburst ever seen from an object outside our Solar System!

Magnetars were predicted by scientists for decades before they were discovered. The first evidence for one was found in 1996. This magnetar, given the name SGR 1806–20, has been studied ever since. It remains one of the most powerful magnetars ever discovered.

A magnetar has a magnetic field about a thousand times stronger than a regular neutron star and a trillion times stronger than Earth's.

Scientists are still trying to figure out what's on the inside of a magnetar. Some ideas include a weird state of matter called a superfluid, where electricity can flow without resistance.

Our planet's magnetic field extends from the interior of Earth and out into space. This magnetic field, shown in this artist's illustration, interacts with the Sun through the solar wind, a stream of charged particles.

MAGNETIC FIELDS ALL AROUND

Magnetars may give magnetic fields a bad rap, but the truth is magnetic fields are all around us and can be really useful. The Earth itself has a magnetic field generated from the giant iron ball at its core that spins as the planet rotates. This is why our compass needles point north and migrating birds and bats have the ability to use their own internal compasses to navigate over long distances.

Even more than that, there is a magnetic field whenever a charged particle, like an electron or proton, moves around, and they go hand in hand with electricity. This means that humans use things every day, from toasters to medical devices, that produce magnetic fields. Scientists have figured out how much magnetism is healthy for us, and these common uses of electricity and magnetism are supersafe. For proof, look at the numbers. Most household items produce tiny fractions of one unit of magnetism called a gauss. An MRI (which stands for *magnetic resonance imaging*) machine generates about thirty thousand gauss, which still won't hurt us. A magnetar across the galaxy? It creates a quadrillion—or a thousand trillion—times stronger magnetic field.

ASASSN-14LI:

TURNED INTO SPAGHETTI BY A BLACK HOLE

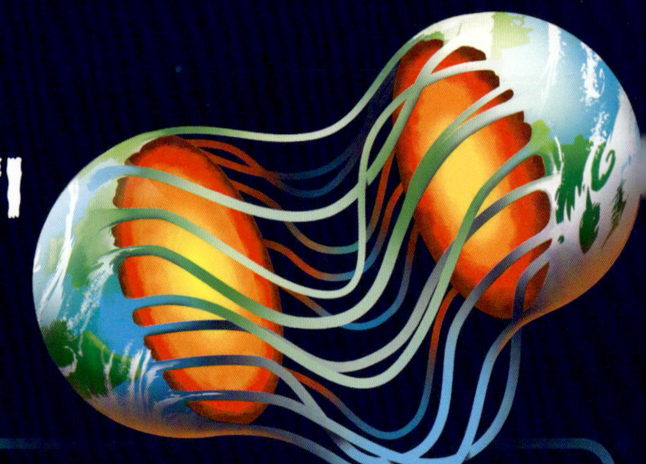

IT'S TIME NOW TO DIVE INTO—not literally—one of the scariest topics in space: black holes. Of course, black holes are on the top of almost any cosmic villain list. You may have heard that nothing, not even light, can escape the clutches of a black hole if it gets too close.

Indeed, these are the beasts of gravity. If you happen to fly a spaceship too close, things get bad pretty quickly. Some black holes have terrorizing whirlpools of dust and gas fueled by friction and can generate blasts of deadly radiation for up to trillions of miles.

Suppose you survive this onslaught; nothing good awaits as you pass over the **event horizon**. This is the line that scientists have calculated is the point of no return. Every black hole has one. As you approach the event horizon, the parts of you nearer to the black hole are pulled up to trillion times more than the opposite end. Soon you are a longer, distorted version of yourself. You are being *spaghettified.*

This artist's illustration shows what happens after a supermassive black hole snacks on a very hefty star. The star was shredded to pieces once it came too close to the black hole, and some of the gas (shown in red) orbited around, and eventually fell into, the black hole. Scientists think the star was about three times the mass of our Sun, making it one of the largest stars ever known to be destroyed in this way.

While no one has ever been to a black hole (or traveled anywhere close to one), scientists are confident that you would be painfully pulled apart by this process, molecule by molecule. Anything that gets trapped by the black hole would be squeezed horizontally and stretched out vertically—ouch!—resembling the famous spaghetti shape. Spaghetti is awesome to eat. It's not awesome to turn into.

Does a black hole really "eat" a star? Here's what happens: 1) A star accidentally falls into an orbit too close to a supermassive black hole. 2) The black hole pulls in the outer part of the star's gas. 3) The star is, sadly, ripped apart. 4) The star's leftovers are pulled into a ring around the hungry black hole, being spaghettified and falling in, creating tons of light and energy in the process.

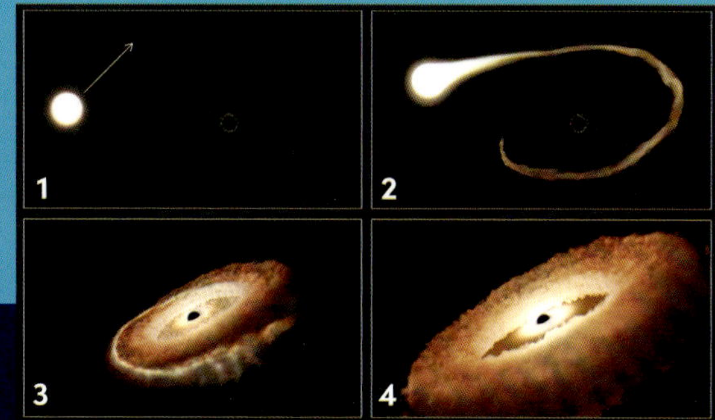

If you think this is a sort of science fiction fairy tale, think again. Astronomers have evidence that death-by-black-hole events are real. Take, for example, the black hole known as ASASSN-14li. (Notice that the first part of its name almost spells the word assassin. Coincidence? Maybe. Or maybe not. ASASSN actually stands for All-Sky Survey for Supernovae, which is a network of telescopes that looks for exploding stars.)

Anyway, this monster of a black hole lives in the center of a galaxy...and it ate a star. First it ripped it apart into shreds, spit some of it back into space, and then consumed the rest. The chunks of the star that fell into the black hole were spaghettified as they went down.

While it would be great to have a live "black hole cam" on all this action, we sadly haven't invented those yet. However, there are telescopes that can pick up clues, and astronomers act like detectives to piece them together. In the case of ASASSN-14li, they analyzed the details of light collected by two X-ray telescopes in space to determine that this black hole, which is safely 290 million light-years away from Earth, devoured a star at least three times as big as our Sun.

That brings us to an important point: How do we find these mysterious and dangerous objects lurking in space? After all, they can't emit light because their intense gravity is so strong. Luckily for us, black holes leave their mark on the space around them. While black holes are completely dark within their event horizons, they also send energy and material outward. This is often in the form of jets or other blasts. Most of these signals are very energetic, so telescopes that see X-ray and gamma ray light are often the best way to find and study them.

You may be wondering what would happen to you once you entered a black hole. Scientists have no idea! Most agree that you wouldn't end up in a different dimension or on the other side of the Universe, even though that's what some people want to believe. However, it's impossible to know for sure given our current information. It's the ultimate unknown ending, so best to avoid it if you'd prefer to eat spaghetti and not become it!

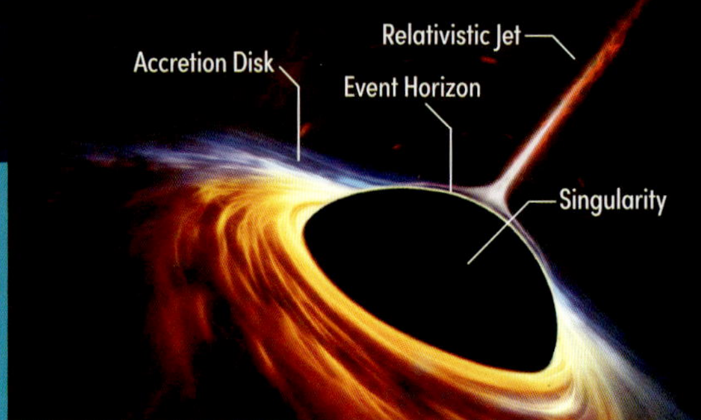

This illustration shows the anatomy of a black hole. There's the event horizon, the line past which there is no return. Black holes send energy and material outward, often in the form of jets. And some black holes have swirling disks of material that surround them.

The nearest black hole (as far as we know) is about fifteen hundred light-years—or about nine trillion miles—away. That's far enough away that the Earth never has to worry about it.

The most distant black hole found to date is about thirteen billion light-years from us. Since scientists think the Big Bang happened about 13.8 billion years ago, that means the Universe had to form these black holes relatively quickly. Scientists are still trying to figure out how that happened.

Black holes come in different sizes. The smallest are **stellar-mass black holes**, which form when a giant star collapses on itself. The largest, called supermassive black holes, are in the centers of most galaxies, including our own. Astronomers think there is a class of black holes in between these two called **intermediate-mass black holes**, but they've been harder to find and pin down.

Our Universe is unimaginably big, so scientists use light-years—the distance that light travels in a year, about six trillion miles—to measure it. Though the center of our Milky Way is twenty-six thousand light-years away, that is relatively close by. One of the most distant galaxies found is believed to be about 13.5 billion light-years away.

WHAT IS A LIGHT-YEAR?

When we talk about distances in space, one unit that is often used in a light-year. This can be one of those confusing terms because it sounds like it measures time and not distance! However, a light-year is definitely a unit of distance. It is the length that light travels over one year. Since light always travels at the same very high speed (about 186,200 miles per second, or about 670 million miles per hour), astronomers can calculate how much distance it would cover in one year. This turns out to be approximately 6,000,000,000,000, or six trillion, miles.

You may have noticed that if you are using miles to describe that distance, that's a lot of zeros to keep track of. The nearest star to the Sun is about four light-years away, which is about twenty-four trillion miles, so that's not too bad. The center of our Milky Way is about twenty-six thousand light-years away, which is about 170,000,000,000,000,000 (170 quadrillion) miles. So when we talk about things outside of the Milky Way that are thousands or even millions of times farther away, it's much easier to use light-years than to write out all those zeros.

The Milky Way is 600,000,000,000,000,000 (six hundred quadrillion) miles across. Can you calculate how many light-years that is?

GAMMA RAY BURSTS:

THE HULK MEETS THE DEATH STAR

CHANCES ARE YOU'VE PROBABLY heard about the Death Star (or its relatives) from the *Star Wars* series. In case you haven't, it's a moon-sized weapon that can shoot beams of energy strong enough to destroy entire planets. The Death Star is what keeps Luke, Leia, Yoda, Rey, and others up at night.

Star Wars is fictional, but what if a planet-killing beam existed? Or worse yet, what about a deadly beam that could reach across an entire galaxy?

Introducing gamma ray bursts, or GRBs for short. Astronomers have known about these killer death rays in space for decades. GRBs are intensely strong blasts of, well, gamma rays that travel across the Universe. If gamma rays sound familiar, you might have heard about them from another science fiction legend: the Hulk. According to that storyline, Dr. Bruce Banner was accidentally shot with gamma rays during an experiment gone wrong, and the rest is superhero history.

When a very massive star explodes, it produces a searing blast of gamma rays in what's called a gamma ray burst (GRB) in this artist's illustration. We can detect such GRBs when this type of high-energy event is pointed in the direction of Earth.

Gamma rays are very real. They are,

in fact, the most energetic and potentially deadly form of light in the known Universe. Luckily for us on the surface of Earth, our atmosphere acts as a sort of force field against most high-energy light like X-rays and gamma rays, keeping us relatively safe.

If GRBs ever went off in our galactic neck of the woods, they could do some major damage. The good news is that because of where Earth is situated in the Universe, scientists don't believe we are close to any sources of GRBs, black holes with death ray potential.

As you travel up to higher and higher altitudes on Earth, our protective atmosphere eventually thins out and fades to the blackness of space. Astronauts on board the International Space Station captured this image of the top of Earth's atmosphere with only a hint of our Moon emerging from the halo.

Black holes create GRBs when a massive star—perhaps forty times the mass of our Sun—collapses on itself. These are no ordinary stars. They are not even ordinary big stars. The stars that can create a black hole in space are colossal. The battle between the incredible gravity pulling on the stars versus the immense pressure pushing outwards in the star ends in an epic energetic event.

During such an explosion, the energy from the star's collapse is directed into a thin beam of energy and particles. We can only see such an event on Earth if the GRB is pointed roughly in our direction. If it's not, chances are our telescopes won't even pick the signal up.

Scientists think that there are two kinds of GRBs: some that last less than two seconds and some that glow between two and thirty seconds. Both probably represent the birth of a black hole, but scientists think the black holes in the longer category are usually formed when two neutron stars collide.

On October 9, 2022, telescopes in orbit around the Earth detected an extremely powerful gamma ray signal. It was dubbed GRB 221009A. (In this case, *22* for the year, *10* for the month, and *09* for the day.)

Scientists quickly determined that this was the brightest and most powerful gamma-ray burst ever seen. This led to nicknaming GRB 221009A the "BOAT" (brightest of all time), a play off the phrase GOAT (greatest of all time), which is often used for awesome sports figures. Who knows if another BOAT will come along in the near future? If so, we may need a new storyline from science fiction in addition to the Hulk and *Star Wars* to help understand it.

The Hubble Space Telescope captured the cooler infrared afterglow of the gamma-ray burst called GRB 221009A, along with the galaxy it lives in, within two months of the eruption in 2022.

Gamma-ray bursts were discovered accidentally in the 1960s by U.S. satellites looking for Soviet testing of nuclear weapons. Scientists soon realized the signals they were picking up were coming from deep space—not anywhere near Earth—and the hunt was on for figuring out what they were.

Gamma-ray bursts can release more energy in ten seconds than our Sun will release in its entire ten-billion-year lifetime.

The most distant gamma-ray burst found yet was 12.8 billion light-years away. Consider that the entire Universe is about 13.8 billion years old, so this GRB went off "only" a little less than a billion years after the Big Bang.

NUCLEAR FUSION VS. GRAVITY

Stars often get poetic treatment in music and literature: *the stars in your eyes, a sky full of stars, as pretty as all the stars*, etc. But these cosmic wonders have a less pretty underbelly. Stars are, in fact, the sites of long battles that can rage for millions or even billions of years.

On one side, there is the engine that powers the star called nuclear fusion. This is an atomic process where atoms smash together and release light and energy. On the other side of the battle, there's gravity continuously trying to smash the star to its core. It's push versus pull, or nuclear fusion versus gravity. As long as the star has atoms and molecules to fuel the nuclear fusion engine, it will survive. Once that fuel runs out though, gravity wins. If the star is big enough, it will explode as a supernova and maybe birth a black hole. If it is not big enough, it will turn into a dim and colder version of what it once was. This is called a white dwarf, or a dead star that eventually cools until it fades from view.

Thousands of stars light up a portion of the sky in this Hubble Space Telescope image of the star cluster called Liller 1. Less than 30,000 light-years from Earth towards the center of our Milky Way, some of these stars formed over eleven or even twelve billion years ago, while some are younger at only one or two billion years. All are fighting the good fight of push versus pull, nuclear fusion versus gravity.

HYPERVELOCITY STARS:

TRAPPED ON A RUNAWAY STELLAR TRAIN

HAVE YOU EVER WATCHED a movie or TV show that has a runaway train because someone cut the brake lines? Hollywood likes when objects speed out of control because it's terrifying. The universe, however, is way ahead of movies and TV. Want proof? Look to the **runaway stars**!

These are not stars that sneak out after curfew. These are stars that are hurtling through space at millions of miles per hour—a thousand times faster than a speeding bullet. And these stars aren't headed for a blown-up bridge or wrecked train tracks. Instead, they are speeding toward the edge of the galaxy, where they will eventually fling themselves into the emptiness of **intergalactic** space.

What launches a lone star into the blackness of space all by itself? Astronomers think this one-way ticket out of the galaxy comes when stars wander too close to something humongous

A runaway star can plow through space and, like a snowplow, pile up interstellar material in front of it. This image from the GALEX ultraviolet satellite shows such a star, CW Leo, as it hurtles through space at over two hundred thousand miles per hour. The star is shedding its outer layers, seen in the center of this image as a bright blue circular blob. The star is moving from right to left and traveling so quickly that it has formed a semicircular shock in front of itself, like what happens when a boat is moving through water.

The star Zeta Ophiuchi was once in close orbit with another star before being ejected when its companion was destroyed in a supernova explosion. You can see a beautiful shock wave flowing away from the star's surface and slamming into gas in its path.

that weighs a lot, like a supermassive black hole or a very large explosion like a supernova.

Stars often go through space with partners. While our Sun is a solo star, about 85 percent of all stars travel in pairs or even trios. If a pair wanders too close to the edge of the black hole, the pair is broken apart. One of the stars gets pulled into the black hole, gone forever. The other can be flung away from the black hole as if launched by an enormous slingshot.

If you were on one of these stars, it would be a rough ride. Any planets could be long gone, left behind in your stellar wake as you raced through deep space. Instead of a night sky filled with the lights of the Milky Way, you would see mostly darkness.

Rest assured that our Sun and Earth are really far away from the Milky Way's supermassive black hole. However, there are many other stars that aren't. Most of the time, these nearby stars orbit around this cosmic sinkhole without a problem. However, if their orbit gets altered even a little bit, it can put them on a path of peril toward the black hole's gravitational trap.

Astronomers first predicted the existence of runaway stars in 1988 using calculations and models. Seventeen years later, scientists using a telescope in Arizona found evidence of one of these runaway stars racing straight out of the Milky Way! We've been watching them depart ever since.

Today we know that there are hundreds of these lone stars hightailing it across the galaxy, many of them destined for an eternity of banishment into the blackness of intergalactic space. Astronomers know that the fastest of these runaway stars will be ejected from their home galaxy, never to return.

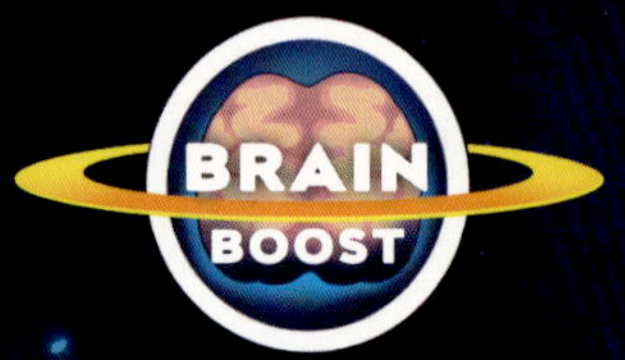

One of the fastest runaway stars found yet is known as HE 0437–5439, which travels at a speed of about 1.6 million miles per hour. Astronomers using the Hubble Space Telescope discovered this speedster heading away from the Milky Way's core and toward intergalactic space.

Astronomers think that these stellar outcasts are pretty rare, with about one in one hundred million stars being booted out of the Milky Way by our supermassive black hole. Since scientists estimate there are about a hundred billion stars in our galaxy, they expect there are about a thousand runaway stars total.

Studying these hypervelocity stars gives information about the landscape of the Milky Way itself, something about how often material gets too close to its black hole, and even how much dark matter lurks throughout the galaxy.

A star called HE 0437–5439 was tossed out of our Milky Way galaxy with a lot of speed, like a superpowerful baseball batter hitting a home run right out of the park. HE 0437–5439 is now rocketing its way through the distant outskirts of our galaxy at over 1.5 million miles an hour! This artist's illustration shows a large blue star's path as it exited the center of our galaxy and up into intergalactic space.

LEARN LIFT: STARLIGHT, STAR BRIGHT

Next to the Sun and Moon, stars are the objects from space that we can see from Earth most easily. While they look pretty much the same from our vantage point on the ground, they are a diverse group. The most important differences in stars come in their colors, which indicate their temperatures and sizes.

There are so many things to learn about stars. Here are some highlights:

Stars have finite lives. They are born, live for a certain amount of time, and ultimately die.

The length of time each star lives and how it dies are dictated largely by how big, or massive, they are. The most massive stars tend to live the shortest lives, and the smallest, dimmest stars tend to live the longest.

Thousands of stars of different sizes and ages are all shown in this gorgeous star cluster called Liller 1, located about thirty thousand light-years away. This image was captured with the Hubble Space Telescope. It took about eleven billion years to form these stars.

Many of the elements necessary for life as we know it on Earth (including the oxygen we breathe, the iron in our blood, and calcium in our bones) were forged inside of stars long ago.

Most of the stars we can see in our night sky are much bigger and brighter than our Sun, and we can see a couple thousand stars at most from the darkest night-sky locations on our planet. In most other locations, such as in suburbs and cities, we see much fewer stars because human-made light from the ground (known as **light pollution**) usually interferes.

We often think of the color red as representing something hot. In fact, red represents the lowest temperature that a heated object can glow in visible light. As an object gets hotter, its color changes to white and finally to blue. So, in the night sky, any red stars you see are the coolest stars, while the blue stars are the hottest.

3C321:

WHEN ONE GALAXY ATTACKS ANOTHER

IN STAR WARS, the Empire's weapons always seem to get bigger and more destructive. But what could be scarier than the power of gamma-ray bursts? Try a giant galaxy, which scientists have nicknamed a "death star galaxy," that is firing an enormous deadly beam and blasting an entire neighbor galaxy.

In the system called 3C321, there are two galaxies orbiting each other. They are not on a collision course, but they are locked together through gravity, the way the Moon orbits the Earth. Both galaxies in 3C321 have supermassive black holes at their centers.

The bigger galaxy has unleashed a gigantic jet of particles and radiation out into space. This galactic death ray has likely been firing for millions of years, but now the smaller galaxy has swung into the beam's path during its orbit!

Nicknamed the "death star galaxy" by scientists, 3C321 is a galaxy with a powerful supermassive black hole at its core that produces incredible amounts of high-energy radiation. The jet from the galaxy is hitting another smaller galaxy nearby. Over a billion light-years away from Earth, we are fortunately at a very safe distance from the death star galaxy.

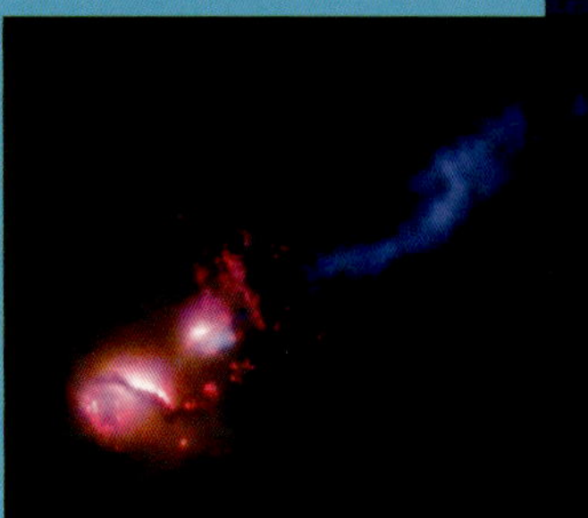

This real-life Death Star beam is traveling at hundreds of millions of miles per hour, or nearly the speed of light, carrying very high amounts of radiation, mainly X-rays and gamma rays. It may ravage the atmospheres of any planet in the smaller galaxy.

How does a black hole launch this galaxy-sized death ray? Black holes are often misrepresented as a sort of cosmic vacuum cleaner, sucking up anything that comes remotely near them, but that's not the case.

This artist's illustration of the death star galaxy 3C321 shows the main larger galaxy and its smaller companion galaxy. A jet of particles generated by the supermassive black hole at the center of the larger galaxy reaches out to strike the smaller. The jet is then disrupted and deflected by the impact.

Instead, black holes have the most impact on their surroundings in what doesn't fall into the event horizon, that is, the point of no return. Imagine that black holes are like cosmic engines, powered by their superstrong gravity. Black holes can collect huge disks of gas and dust swirling around them. When some of this material falls toward the event horizon, it's redirected from the black hole outward into space as narrow beams. Astronomers have seen these beams from black holes throughout the Universe. Some of the jets are enormous, stretching for hundreds of thousands of light-years—enough to reach another galaxy!

This is what is happening in 3C321. The damage to the smaller galaxy from its larger galactic neighbor is particularly catastrophic because these two galaxies are very close to each other, at least in space terms. At a distance of about twenty thousand light-years, the two black holes and their galaxies are about the same distance from each other as the Earth and the center of the Milky Way. While this might seem like a long way, it is point-blank range for this galactic attack.

The image at the start of this chapter, made of light collected by several telescopes, shows the beam generated from one galaxy hitting the other and then deflecting off to the side like water from a hose. Astronomers were surprised to see this kind of galaxy-on-galaxy assault when they discovered it, but they don't think it's all doom and gloom for the smaller galaxy in

3C321. The death ray from the larger galaxy may cause major damage in certain parts of the smaller galaxy, but it also may ignite new generation of stars and planets after the initial wave of destruction is over.

So while getting punched by your galactic neighbor might be tough, the little galaxy in 3C321 may claim victory some day in the future.

The Event Horizon Telescope, a network of radio antennae around the globe, captured our first image of a black hole's event horizon. It's like looking at the shadow, or the silhouette, of the black hole. This black hole is located in Messier 87, a galaxy about sixty million light-years away.

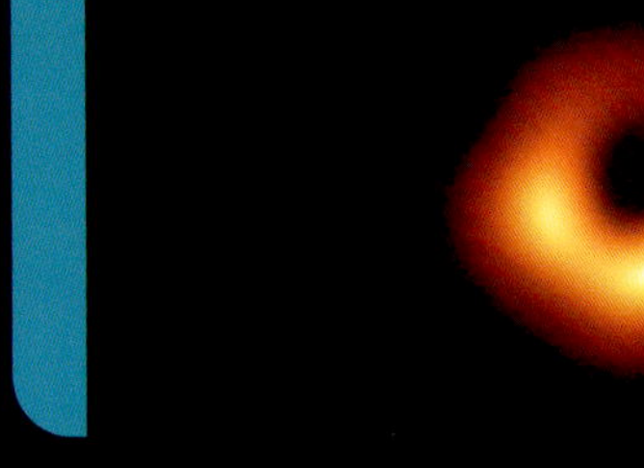

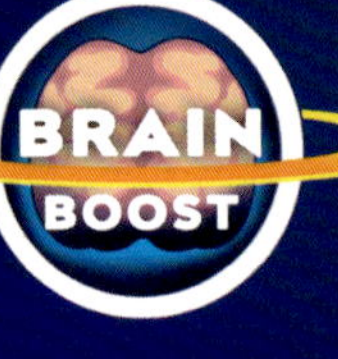

Albert Einstein's famous theory of general relativity allowed for the possibility of black holes, but Einstein himself did not think they could be real. Astronomers, however, have been studying black holes using telescopes since the 1970s. Unfortunately, Einstein died in 1955, so he never got to see proof of his theory.

Today, scientists have collected lots of data on black holes, including generating the first image of a black hole from a network of radio telescopes called the Event Horizon Telescope in 2019.

While astronomers still cannot see inside a black hole, using current telescopes, they can observe to their edges in some cases and track what kind of impact a black hole has on its neighbors. Astronomers are designing and building new telescopes that will be able to see even more details about black holes in the future.

THE DIFFERENT SIZES OF BLACK HOLES

There are three main categories of black holes based on how much mass, or amount of matter, they have. The most common kind of black hole is known as the *stellar-mass* variety. These are black holes that are created when a gigantic star runs out of fuel and collapses on itself. While there are no stellar-mass black holes close to us, astronomers know there are about twenty in the Milky Way galaxy. Our Sun will not become a black hole—it does not have enough mass.

The next type is the supermassive black hole. These are the giant black holes found in the centers of galaxies, including our Milky Way. They can have millions or even billions of times the mass of the Sun. Scientists aren't exactly sure how supermassive black holes form, but it likely involves merging with smaller black holes and pulling in lots of dust and gas in a process that takes billions of years.

The last category is known as intermediate-mass black holes. As the name may suggest, these are black holes that fall in between stellar-mass and supermassive, in terms of size. Over the years, astronomers have found evidence of this kind of black hole, but the jury is still out on whether they exist and what their role is in the cosmic ecosystem.

Scientists think that black holes might come in three sizes: small, medium, and large. Stellar-mass black holes are the smallest. Supermassive black holes, like this illustration here shows, are the largest. There is possibly a third size of black holes called intermediate-mass, but they have been very difficult to find.

DARK MATTER:
CONTROLLED BY SOMETHING YOU CAN'T SEE

Everything we can observe—humans, trees, Earth, planets, stars, galaxies—is made up of matter, meaning it has substance, or mass, and takes up space. As you've learned throughout this book, scientists have become better and better at seeing matter and light in space.

In the 1930s, Fritz Zwicky looked at the Coma galaxy cluster and determined that the galaxies within it were moving so fast, they should have been flung out into space. But they weren't. He argued that something was holding them back.

About four decades later, astronomer Vera Rubin was studying spiral galaxies and saw something too. She noticed the stars around the edges of the spiral galaxies moved in a way that didn't make sense. Even if she combined the mass of all the stars, dust, and gas, it didn't add up. Again, something invisible with a gravitational pull was out there.

The Coma cluster contains thousands of galaxies over three hundred million light-years away from us, shown here in optical light from the Sloan Digital Sky Survey. The two large white "eyes" in this image are ancient elliptical galaxies that are likely the result of previous massive collisions of smaller galaxy clusters. There might be a lot of space in space, but there's also plenty of stuff to collide with!

Rubin's work eventually convinced other scientists that something was out there that they could not see. It was dubbed dark matter because it doesn't interact with light in any way that we know about.

Here's what we do know: Dark matter is an invisible glue that makes up and controls a bulk of the Universe's matter. It also isn't made of atoms and doesn't carry any electrical charge. This means it isn't like any of the rest of the "normal" matter we see here on Earth, in stars, or in galaxies.

Since Vera Rubin's work, other astronomers have found evidence for dark matter. Take, for instance, galaxy clusters. These are gargantuan structures that have hundreds or even thousands of galaxies, all connected by gravity. On top of that, they are sitting in an enormous pool of superheated gas. Even though this gas is very thin, it has more mass than all of these galaxies put together. But, if you add up all the superheated gas and all the galaxies in a galaxy cluster, you *still* don't have enough mass to explain how these things are held together. It does make sense, however, if there is dark matter that we can't see invisibly tipping the scales.

Nicknamed the "bullet cluster" because of its shape, this massive collision of two large galaxy clusters is billions of light-years away. This image shows hot gas, or most of the normal matter in the cluster, as two pink clumps. The galaxies themselves are in orange and white. The blue clumps demonstrate that most of the mass is separate from the normal (pink) matter.

Need more proof? In 2006, astronomers using telescopes in space and on the ground looked at a system called 1E 0657+56. It's better known by its nickname, the "Bullet Cluster." It's where two gigantic galaxy clusters smacked into each other.

By looking at X-ray light, astronomers could trace where most of the regular matter was in the colliding galaxy clusters. Using a special technique, they figured out where gravity was the strongest, and it wasn't in the same place. So there was gravity...but no stuff. In that case, what is responsible for the gravity? Astronomers think it's dark matter.

There are a bunch of ideas being kicked around about what dark matter is, including the idea that there are very teeny tiny particles that behave differently than the usual electrons, protons, and neutrons. Scientists have built gigantic detectors deep underground—far away from the other particles we usually find on our planet's surface—to look for the signals of these kinds of particles. That search is still ongoing. Astronomers are also looking for more clues to what dark matter might be. They are studying everything from really energetic particles that whiz through space to some of the fingerprints of light left behind by the Big Bang.

Astronomers call the DF2 galaxy shown here a see-through galaxy because they can almost see clearly through it. The DF2 galaxy has 1/200th the number of stars as the Milky Way and only about 1/400th the amount of dark matter.

An invisible force that we can't directly see and don't understand, controlling things throughout the Universe? Scary! Dark matter is one of the biggest mysteries in the Universe and definitely one of those that can keep you up at night if you like to know what's out there.

BRAIN BOOST

Scientists estimate that 85 percent of all of the matter in the Universe is dark matter. That means there is about six times more dark matter than there is "normal" matter, or stuff, in the Universe.

Scientists love acronyms. One of the leading ideas for what dark matter could be is something scientists call **WIMPs**, which stands for weakly interacting massive particles. As the name suggests, these would be relatively big particles that don't interact with anything else.

A small fraction of scientists do not think dark matter is real. They think that instead there is something about gravity that we do not understand. Since this would contradict Einstein and most everything else modern astrophysics is based upon, the vast majority of scientists think dark matter is real.

NOBEL PRIZES

One of the most important awards that scientists receive is the Nobel Prize. These are given out in several categories, including physics.

Only five women have been given the physics award in the entire history of the Nobel Prize, which started in 1901. Many people argue that this is not because there haven't been deserving women, but often because women—and other underrepresented groups of people—don't get the recognition they deserve. Search for Rosalind Franklin and Chien-Shiung Wu for examples of incredible scientists. People are working hard to make the Nobel and other prizes fairer, but there's more work to be done.

Vera Rubin was one of those scientists who should have been given the Nobel for her work on dark matter, which upended our understanding of what the Universe is made of. However, she passed away in 2016, and the Nobel Prize is only awarded to people who are still living, so we will never have the chance to right that wrong. However, there is a consolation prize. A new giant telescope in Chile will survey the Universe and help investigate dark matter has been named the Vera C. Rubin Observatory, providing a huge honor for an astronomer often overlooked during her lifetime.

The Vera C. Rubin Observatory in Chile, which began operations in 2025, will conduct a unique decade-long survey of the sky in optical light to help scientists better understand the nature of dark matter, the structure of our galaxy, and catalog our own Solar System.

DARK ENERGY:

ULTIMATE EMPTINESS AT THE UNIVERSE'S END

LOOKING UP AT THE NIGHT SKY from Earth, all we see is a flat picture. Some stars are bigger and brighter than others, but that doesn't tell you how far away anything is. What if one star was just brighter?

It took a long time and a lot of very hardworking and smart people, but astronomers finally came up with ways to tell how far away things were. By the end of the 1920s, most astronomers were convinced that the Universe was expanding like a giant balloon.

This was a tough idea to believe. After all, how and why would the universe expand? This led to the idea of the Big Bang, or the explanation of how the Universe started in an explosion 13.8 billion years ago. We're still moving away from that explosion all these years later as the Universe expands outward.

Our Universe is unimaginably big! This image is not of a population of stars but rather of a massive group of thousands of galaxies 240 million light-years from us. The galaxies close to us look like larger cotton balls of bright light, but hundreds of thousands of much smaller galaxies are visible as pinpricks of light. Dark energy is slowly pushing all of these galaxies apart.

Then came the late 1990s with another bombshell of a discovery: Astronomers found that the Universe was expanding much faster than they would have ever expected. Nothing they knew could explain it. Then they did what scientists often do when they don't understand something:

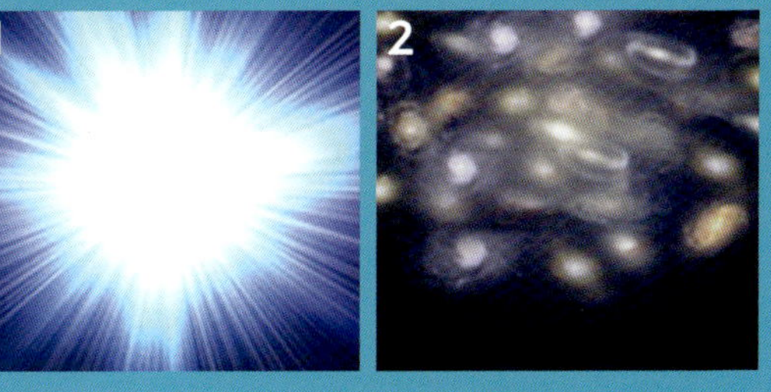

This sequence of illustrations shows the expansion of our Universe from the Big Bang, shown as a flash of light (1), immediately followed by the rapid expansion of the Universe (2). The expansion then slows down because of the gravitational attraction of the matter in our Universe (3). As the Universe continues to expand, dark energy becomes important, causing the expansion to accelerate until, far in the future, galaxies will be so spread out we won't be able to see any other galaxies at all (4).

They gave it a name. They called this mystery dark energy.

There are so many things we don't know about dark energy. Most importantly, we simply do not know what it is. What we do know is that dark energy is making the expansion of the Universe get faster and faster. Beyond that, scientists have a bunch of ideas. Some think it's a kind of reverse gravity that pushes everything apart instead of pulling things closer together. Others suggest that it's an energy that we've never encountered before. Or maybe gravity works differently over large distances?

Because this is science, we need to have data that we can test over and over again to prove what dark energy really is. And scientists are planning to do just that. In 2023, the European Space Agency launched a mission called Euclid that will make the best 3D map ever of the Universe. This will let scientists measure more exactly where the Universe is expanding and by how much, which will give them more clues to what dark energy is.

NASA is also trying to get to the bottom of the dark energy mystery. A new mission, scheduled to launch in the late 2020s, called the Nancy Grace Roman Space Telescope, will search for matter across the Universe, including dark matter. Again, this will help scientists figure out how dark energy has behaved during the billions of years the Universe has been around, and this will narrow down the options of what it could be.

It might take decades for humans to figure out what dark energy is. That may be okay because we have a long time before it starts to affect us. You may have noticed that around here, the rules of gravity stick, including in our day-to-day lives. If you throw something up, it falls to the ground. The Moon orbits the Earth, and the Earth goes around the Sun, all thanks to gravity. Even throughout the Milky Way and beyond, gravity is reliable.

So what's up with dark energy? It seems that it is something that only has an effect over millions or even billions of light-years. For anything on a smaller scale, dark energy's influence is too tiny to notice.

A Type 1a supernova happens when a white dwarf star pulls material from, or merges with, a nearby companion star, triggering a violent explosion. The white dwarf star is obliterated, sending its debris hurtling into space. This type of supernova is used as a cosmic mile marker because its brightness is so reliable. This image is of a Type 1a supernova, the Tycho Supernova Remnant.

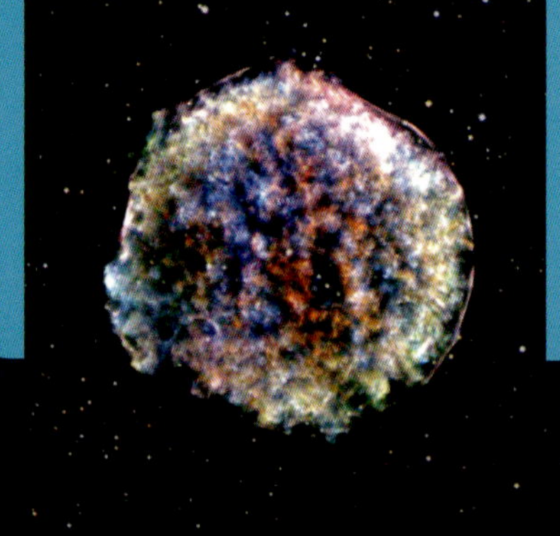

If we can't see it or feel it, should we even care about dark energy? Yes, because as long as dark energy is driving the Universe apart, its fate isn't pretty. According to theoretical physicists, scientists who use math and data to explain the Universe, dark energy could change and get stronger over time. If this is the case, this will cause the Universe to end in a cosmic apocalypse that scientists call the Big Rip.

In the Big Rip, dark energy will take over all the other forces that we know and live with today, including gravity. When and if this happens, everything will come apart. First galaxies will come apart, then the planetary systems, then atoms will dissolve, and even space itself will tear. There will be absolutely nothing left.

While this may be one of the freakiest ideas ever, it's definitely one we can all put far in the back of our minds. Because if dark energy is controlling our fate, the Big Rip couldn't and wouldn't happen for hundreds of billions of years. Maybe, by then, humanity's future descendants will have come up with a remedy.

Dark energy was discovered in 1998 by two separate teams of astronomers studying **Type 1a supernovas**. If one supernova is dimmer, it's farther away, which helps scientists measure large distances in space. These special supernovas revealed the Universe was moving apart faster than anyone expected.

Since then, astronomers have tested galaxy clusters and other objects scattered throughout the Universe to look for dark energy. They have found the same effect of dark energy as the supernova teams did.

Dark energy is a great example of why people do science! You can discover things that you don't expect and can't explain. The fun is trying to find the answers and then offer proof to convince the rest of the world.

THE INTERNATIONAL SPACE STATION AND BEYOND

Since 2000, there have been humans living in space, about 250 miles above Earth, aboard the International Space Station (ISS). Several nations, including the United States, Russia, and Japan, worked together to help build the ISS over ten years. To construct the ISS, different sections, or modules, were launched separately and then assembled while in space. Sometimes one country built the module, and another launched it. The translators on this project were very important!

A lot has changed in human spaceflight in recent years, especially on the American side. When the ISS first started, NASA used the Space Shuttle to bring astronauts and equipment into orbit. Today, the Space Shuttle is retired, and most of these trips are conducted by private space companies, including SpaceX and others. NASA's plans are no longer focused on the ISS because the goal is to build human settlements even farther out. They hope to create a human outpost on the Moon next, and then one day do the same on Mars.

The Universe may be enormous, but these small steps are important—and hard!—for sending humans out into space.

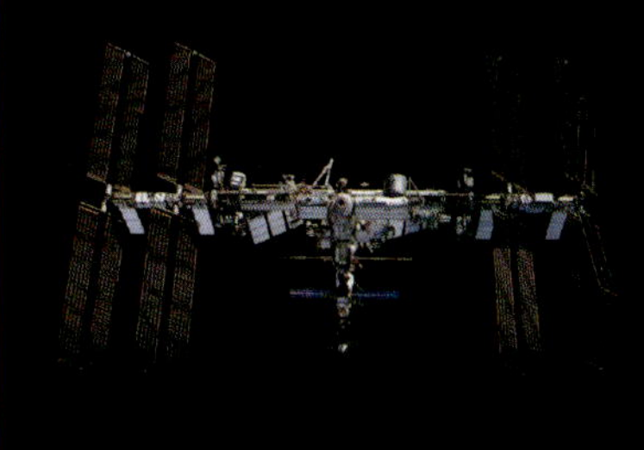

What's it like to live on the International Space Station (ISS)? The ISS is larger than a six-bedroom house, with a lot of complex cables, computers, and parts strewn throughout. It comes with sleeping quarters for six astronauts, two very small bathrooms, a gym (to keep muscles, bones, and lungs all stay healthy), and a 360-degree view window called the cupola.

GLOSSARY

AMMONIA: a gas found in the atmospheres of many planets and moons that is dangerous and potentially deadly to humans

ANTICYCLONE: a storm that rotates in a counterclockwise direction, as opposed to most hurricanes or cyclones

ASTERISM: a familiar star pattern, such as the Big Dipper, that is not an official constellation

ASTROBIOLOGY: the study of how life emerged on Earth and the search for where it could exist elsewhere

AURORAS: often referred to as the Northern Lights, a colorful display in the skies around either pole

BLACK HOLE: an object that has so much matter concentrated in such a relatively small space that nothing can escape its gravity within a certain distance

CONSTELLATION: an officially recognized group of stars in a pattern that is often tied to a story or myth

CRYOVOLCANO: a volcano that erupts with ice, cold water, and other materials instead of molten rock and ash

DARK ENERGY: a mysterious source of energy that scientists think explains the acceleration of the expansion of the Universe

DARK MATTER: an invisible glue that makes up, and controls, a bulk of the Universe's matter

DEEP SPACE: generally refers to anything beyond the Moon

DWARF PLANET: an object that is similar to planets but smaller

EVENT HORIZON: the boundary around a black hole beyond which nothing, not even light, can escape

EXOMOON: a moon that orbits a planet outside of our Solar System

EXOPLANET: a planet that orbits a star other than our Sun

FAILED STAR: an object that is big enough to be a star but not big enough to sustain nuclear fusion

FROSTBITE: when the tissue beneath the outer layer of skin freezes

GALAXY: a system of stars, dust, black holes, and more all held together by gravity

GALAXY CLUSTER: a collection of hundreds or thousands of galaxies usually immersed in superheated gas

GAS GIANT: a large planet that is mostly hydrogen or helium, including Jupiter and Saturn

GEYSER: a type of hot spring that erupts

GOLDILOCKS ZONE: see **habitable zone**

GRAVITATIONAL WAVES: invisible ripples in space that move at the speed of light

GREENHOUSE EFFECT: when gases in a planet's atmosphere trap light and heat, causing the planet to warm

HABITABLE ZONE: the distance from a star in which the temperature allows for liquid water to exist on a planet

HOT JUPITER: a planet roughly as large as Jupiter or larger that orbits very close to a star

HYDROCARBONS: a class of chemicals made of only hydrogen and carbon, often found in fossil fuels

HYDROTHERMAL VENT: an opening on the floor of an ocean or sea where heat and material escape

ICE GIANT: a huge planet made up mainly of ammonia, water, and methane, like Neptune and Uranus

INTERGALACTIC: describes a region in between or around galaxies

INTERSTELLAR: a region in between or around stars

INTERMEDIATE-MASS BLACK HOLE: a not-yet-confirmed class of black hole with masses between about a hundred to a hundred thousand times that of the Sun

JET: a beam of energy and particles fired out of black holes, stars, or other objects in space

LIGHT-YEAR: the distance that light travels in a year, about 6 trillion miles

LIGHT POLLUTION: human-made light that interferes with natural light, particularly at night

MAGNETAR: a superdense dead star with extremely powerful magnetic fields

MATTER: anything that has weight and takes up space

NEUTRON STAR: the dense core left behind after a giant star collapses

NUCLEAR FUSION: when two nuclei, or dense centers, of atoms smash together and release energy

pH LEVEL: a measure of the acidity of a liquid containing water

PLANET: a roughly spherical object in orbit around a star

PULSAR: a dead, dense leftover core of a star that gives off pulses of light as it spins

RADIAL VELOCITY: a measurement that can reveal the existence of a planet through the tiny wobble of a star

RUNAWAY STAR: also known as a hypervelocity star, it moves so fast it can escape a galaxy's gravity

SILICATE: a compound that combines includes silicon, oxygen, and other elements, like magnesium and iron

SODA LAKE: a lake that has high levels of dissolved sodium and carbonate, making it similar to baking soda

SPIRAL GALAXY: a type of galaxy that resembles a pinwheel, with long sweeping arms

STELLAR-MASS BLACK HOLE: a black hole with between five and ten times the mass of the Sun

SUPER EARTH: a type of planet outside our Solar System larger than Earth but smaller than Uranus or Neptune

SUPERMASSIVE BLACK HOLE: gigantic black holes with millions or billions more mass than the Sun, typically found in the centers of galaxies

SUPERNOVA: an explosion of a massive star that often sends its outer layers out into interstellar space

TRANSIT: a dip or decrease in a star's light when a planet or other body passes in front of the star

TECTONIC PLATES: larger pieces, or plates, that make up the crust of a planet or moon

TERMINATOR: a line on a planet or moon that marks the boundary between daytime and nighttime

TYPE 1A SUPERNOVA: a special type of exploded star that always has the same brightness

WINDCHILL: a term used for the feeling on human skin of the combination of cold temperatures and high winds

WIMPS: an acronym for *weakly interacting massive particles*, one proposed explanation for dark matter

ZOMBIE PLANET: a planet that has somehow survived a catastrophic event like a supernova explosion

Dr. Kimberly K. Arcand was working in biology when she was hired for NASA's Chandra X-ray Observatory in 1998. Since she always wanted to be an astronaut when she was little, this opportunity brought Kim close to the cosmos but without the long-distance commute. Today, Kim uses data to tell stories about science, whether in the form of a 3D print of an exploded star, a virtual reality application of a pulsar, or the sound of a black hole. Outside of work, Kim likes to hang out with her family, including her husband, two grown-up children, her parents, two dogs, and a cat. She's a big fan of nature and loves to be outside gardening, walking, and looking for her next rainbow.

Megan Watzke has been interested in space and exploration since she can remember. She earned her degree in astrophysics in college, even though she wasn't the best student in her classes. After college, she found a graduate program that taught people with science backgrounds how to write. That led her to current career as a science writer, where she loves making science understandable for everyone, including kids. Megan has learned a lot about what is creepy and freaky from her own four kids, whom she and her wife live with in Seattle, WA.

Robert Ball has worked in the creative sector for over thirty years. He started as a graphic artist in the games industry, became an art director for a branding agency, and finally chose life as a full-time illustrator in 2014. He has created art for book covers, toy and game packaging, and advertising clients. He grew a loyal fan base through his project for HBO's *Game of Thrones*, in which he illustrated a poster for each episode. He lives in London in a house full of books where he gets under the feet of his partner, Anya, and is a full-time butler to their dog, Conker.

Image Credits

Page 8 Sun & Mercury: NASA/JPL • Page 9 black & white Mercury: NASA/JPL; tardigrade: © Schokraie E, Warnken U, Hotz-Wagenblatt A, Grohme MA, Hengherr S, et al. (2012) • Page 10 Mickey Mouse Mercury: NASA/Johns Hopkins University Applied Physics Laboratory/Carnegie Institute of Washington • Page 12 burnt marshmallow Venus: NASA/JPL-Caltech; acid and lava: NASA/JPL • Page 13 illustration of the layers of Earth and Venus: NASA; Venus volcano: NASA/JPL-Caltech/Peter Rubin • Page 14 illustration of planets and Sun: IAU/Martin Kornmesser • Page 16 Europa: NASA/JPL/University of Arizona. • Page 17 Great Red Spot: NASA/JPL-Caltech/SwRI/MSSS, image processing by Kevin M. Gill • Page 18 Jupiter cyclone: NASA/JPL-Caltech/SwRI/ASI/INAF/JIRAM • Page 19 Jupiter's moons: NASA/JPL/DLR • Page 20 Io volcanic eruption: NASA's Goddard Space Flight Center, courtesy of NASA/JPL/University of Arizona • Page 21 Jupiter's four moons: DasWortgewand • Page 22 Io magma ocean: Mark Garlick/Science Photo Library/Getty Images • Page 23 Kepler exomoon: ESA/Hubble • Page 24 Saturn from Cassini: NASA/JPL-Caltech/Space Science Institute • Page 25 Saturn aurora NASA/JPL/ASI/University of Arizona/University of Leicester • Page 26 Saturn upper atmosphere: NASA/JPL-Caltech/Space Science Institute • Page 27 black and white rings: NASA/JPL; infrared light rings: NASA/JPL-Caltech/University of Virginia • Page 28 3 Titan moons: NASA/JPL/University of Arizona • Page 29 Titan dunes: NASA/JPL-Caltech • Page 30 Titan Saturn: NASA/JPL-Caltech/Space Science Institute • Page 31 Cassini Huygens on Titan: ESA–C. Carreau; Rocky Titan: NASA/University of Arizona • Page 32 Enceladus: NASA/JPL/Space Science Institute • Page 33 Geyser tiger stripes: NASA/JPL-Caltech/Space Science Institute • Page 34 Enceladus Saturn rings: NASA/JPL-Caltech/Space Science Institute • Page 35 Great Salt Lake: Alexander Gerst • Page 36 Neptune ice giant: NASA/JPL • Page 37 Neptune's 14 moons: NASA, ESA, CSA, STScI • Page 38 Neptune Triton: NASA/JPL • Page 39 Pluto: NASA/JPL • Page 42 Poltergeist planet: NASA/JPL • Page 43 pulsar magnetic fields: NASA/CXC/M. Weiss • Page 44 Puppis A: NASA/JPL-Caltech/UCLA • Page 45 observatories across spectrum: NASA/CXC/K. DiVona • Page 46 TrES-2b: JPL • Page 47 Kim hot Jupiter: ESO/M. Kornmesser CC, edited by K. Arcand • Page 48 transit graphic: NASA Ames • Page 49 ground and space telescopes: NASA; observatory images from NASA, ESA (Herschel and Planck), Lavochkin Association (Specktr-R), HESS Collaboration (HESS), Salt Foundation (SALT), Rick Peterson/WMKO (Keck), Gemini Observatory/AURA (Gemini), CARMA team (CARMA), and NRAO/AUI (Greenbank and VLA); background image from NASA • Page 50 HD 189733b: NASA, ESA, M. Kornmesser • Page 51 Hubble Space Telescope: NASA • Page 52 illustration of exoplanets: NASA/W. Stenzel • Page 53 25 hot Jupiters: ESA/Hubble, N. Bartmann • Page 54 55 Cancri e: ESA/Hubble, M. Kornmesser • Page 55 comparison chart: NASA, ESA, CSA, Dani Player (STScI), edited by K. Arcand; boiling hot lava: ESO/L. Calçada • Page 56 Copernicus: NASA/JPL-Caltech, edited by K. Arcand • Page 58 illustration of WASP-12-B: NASA/ESA/G. Bacon • Page 59 Earth and Moon: NASA/JPL; cannibalized by star: NASA/CXC/M. Weiss • Page 60 TESS: ESO/M. Zamani • Page 62 Eta Carinae: NASA, ESA, N. Smith (University of Arizona) and J. Morse (BoldlyGo Institute) • Page 63 supernova explosion: NASA, ESA, A. Riess (STScI/JHU), and SH0ES team • Page 64 illustration of very bright supernova: NASA/CXC/M. Weiss; Cas A: NASA, ESA, CSA, STScI, Danny Milisavljevic (Purdue University), Ilse De Looze (Ghent University), Tea Temim (Princeton University) • Page 66 Crab Nebula: NASA, ESA, CSA, STScI, Tea Temim (Princeton University) • Page 67 stellar corpse: NASA/ J. Hester (ASU), M. Weisskopf (NASA/MSFC) • Page 68 Halley's Comet: NASA • Page 69 Hubble Space Telescope: NASA • Page 70 Sagittarius A*: X-ray: NASA/CXC/UMass/D. Wang et al.; optical: NASA/ESA/STScI/D. Wang et al.; IR: NASA/JPL-Caltech/SSC/S. Stolovy • Page 71 illustration of black hole with hot gas: NASA/CXC/M. Weiss; Milky Way's supermassive black hole: EHT Collaboration • Page 73 galactic pancake: NASA/JPL-Caltech/ESO/R. Hurt • Page 76 Andromeda galaxy: NASA/JPL-Caltech, edited by K. Arcand • Page 77 Milky Way: Benjamin Inouye, cc4.0 • Page 78 clashing galaxies: 1: NASA, ESA, STScI, Julianne Dalcanton (CCA/Flatiron Institute and University of Washington); 2: ESA/Hubble and NASA, J. Dalcanton; 3: NASA/STScI; 4: NASA/STScI, PanSTARRS, Judy Schmidt; illustration of Milky Way: NASA, ESA, Z. Levay and R. van der Marel, STScI, T. Hallas, and A. Mellinger • Page 79 illustration of galaxies: A. Feild (NASA/STScI) • Page 80 magnetar: ESA/ATG medialab • Page 81 illustration of the birth of a magnetar: NASA, ESA, and D. Player (STScI) • Page 82 magnetar flare: NASA's Goddard Space Flight Center Conceptual Image Lab • Page 83 magnetic field: NASA's Goddard Space Flight Center • Page 84 illustration of black hole snacking: NASA/CXC/Melissa Weiss • Page 85 black hole eating star: NASA, ESA, Leah Hustak (STScI) • Page 86 illustration of anatomy of a black hole: ESO, ESA/Hubble, M. Kornmesser/N. Bartmann • Page 87 light-years: NASA, JPL-Caltech, Susan Stolovy (SSC/Caltech), et al. • Page 88 gamma ray burst: NASA/Swift/Cruz deWilde. • Page 89 Earth's atmosphere: NASA/JSC/Gateway to Astronaut Photography of Earth • Page 90 infrared afterglow: NASA, ESA, CSA, STScI, A. Levan (Radboud University); image processing: Gladys Kober; illustration credit: NASA's Goddard Space Flight Center • Page 91 Liller 1: ESA/Hubble and NASA, F. Ferraro • Page 92 CW Leo: NASA/JPL-Caltech • Page 93 Zeta Ophiuchi: X-ray: NASA/CXC/Dublin Institute for Advanced Studies/S. Green et al; infrared: NASA/JPL/Spitzer • Page 94 HE 0437–5439: NASA, ESA, and G. Bacon (STScI) • Page 95 star cluster: ESA/Hubble and NASA, F. Ferraro • Page 96 death star galaxy: X-ray: NASA/CXC/CfA/D.Evans et al.; optical/UV: NASA/STScI; radio: NSF/VLA/CfA/D. Evans et al., STFC/JBO/MERLIN • Page 97 illustration of 3C321: NASA/CXC/M. Weiss • Page 98 Event Horizon Telescope: Event Horizon Telescope Collaboration • Page 99 illustration of supermassive black hole: NASA/JPL-Caltech/GSFC • Page 100 Coma cluster: SDSS • Page 101 bullet cluster: X-ray: NASA/CXC/CfA/M. Markevitch et al.; optical: NASA/STScI; Magellan/University of Arizona/D. Clowe et al.; lensing map: NASA/STScI; ESO WFI; Magellan/University of Arizona/D. Clowe et al. • Page 102 DF2 galaxy: NASA/ESA/STScI • Page 103 Vera C. Rubin Observatory: NOIRLab/NSF/AURA/T. Matsopoulos • Page 104 cotton ball galaxies: ESA/Euclid/Euclid Consortium/NASA, image processing by J.-C. Cuillandre (CEA Paris-Saclay), G. Anselmi; CC BY-SA 3.0 IGO • Page 105 illustration sequence of the Big Bang: NASA/STScI/G. Bacon • Page 106 Type 1a: X-ray: NASA/CXC/RIKEN and GSFC/T. Sato et al; optical: DSS • Page 107 ISS: NASA.